T0192785

Single Flux Quantum Integrated Circuit Design

Gleb Krylov • Eby G. Friedman

Single Flux Quantum Integrated Circuit Design

 Springer

Gleb Krylov
University of Rochester
Rochester, NY, USA

Eby G. Friedman
University of Rochester
Rochester, NY, USA

ISBN 978-3-030-76887-4   ISBN 978-3-030-76885-0   (eBook)
https://doi.org/10.1007/978-3-030-76885-0

© Springer Nature Switzerland AG 2022
This work is subject to copyright. All rights are reserved by the Publisher, whether the whole or part of the material is concerned, specifically the rights of translation, reprinting, reuse of illustrations, recitation, broadcasting, reproduction on microfilms or in any other physical way, and transmission or information storage and retrieval, electronic adaptation, computer software, or by similar or dissimilar methodology now known or hereafter developed.
The use of general descriptive names, registered names, trademarks, service marks, etc. in this publication does not imply, even in the absence of a specific statement, that such names are exempt from the relevant protective laws and regulations and therefore free for general use.
The publisher, the authors, and the editors are safe to assume that the advice and information in this book are believed to be true and accurate at the date of publication. Neither the publisher nor the authors or the editors give a warranty, expressed or implied, with respect to the material contained herein or for any errors or omissions that may have been made. The publisher remains neutral with regard to jurisdictional claims in published maps and institutional affiliations.

This Springer imprint is published by the registered company Springer Nature Switzerland AG
The registered company address is: Gewerbestrasse 11, 6330 Cham, Switzerland

To my mother Marina

*To my grandchildren, Hannah and Michelle,
may your curiosity guide you to new and
exciting experiences*

Preface

Conventional semiconductor-based digital electronics, with complementary metal oxide semiconductor (CMOS) technology as the primary example, has experienced meteoric growth over the past several decades. Large scale computing systems, enabled and driven by technology scaling, have transformed human society and our daily lives. Despite the huge resources and attention directed at this technology, the scaling of semiconductor technology is now approaching fundamental physical limitations, encouraging the development of novel beyond-CMOS emerging technologies to replace or supplement existing electronic systems. One rapidly growing application area for these beyond-CMOS circuits is high efficiency, large scale stationary computing—supercomputers and data centers. A technology particularly appropriate for this important application is superconductive electronics. Superconductive electronic circuits have existed for many decades; only recently, however, have the disadvantages of cryogenic cooling, required by these circuits, become less onerous as compared to the fabulous advantages of superconductive electronics in both performance and energy efficiency.

The primary intention of this book is to provide insight into the development of large scale digital superconductive circuits, with particular emphasis on single flux quantum (SFQ)—a type of superconductive digital logic family based on Josephson junctions. SFQ circuits have been known and used now for several decades. The fabrication capabilities of modern superconductive foundries currently approach a million logic gates per integrated circuit. The complexity of practical systems, however, does not currently exceed several thousand gates. This "design gap" exists primarily due to the lack of efficient computer aided design (CAD) tools, in particular algorithms and methodologies aware of the issues posed by the large scale integration of SFQ circuits. The two primary objectives of this book are therefore, first, to provide physical background and engineering intuition into the operation of small scale (device and gate level) superconductive circuits and, second, to introduce the many significant recent developments in this field enabling the large scale integration of superconductive systems.

This book is based upon the body of research carried out by Gleb Krylov from 2016 to 2021 at the University of Rochester during his doctoral study under the supervision of Professor Eby G. Friedman. It has become apparent to both authors during this period that a large body of work exists in this field which has not been covered by existing monographs. These books primarily target the physics of superconductivity and perhaps some small scale analog circuits. The purpose of this book is to introduce and systematize this material on superconductive electronics into a cohesive whole to support the further development of SFQ-based computing systems. A particular emphasis is placed on the many challenges faced by modern superconductive logic circuits and integrated digital systems.

The organization of this book is consistent with this primary purpose and utilizes a bottom-up approach, moving up the abstraction layer stack. The first seven chapters introduce and provide essential background into superconductive electronics. The first two chapters provide a brief background into the physics of superconductivity and the operation of common superconductive devices. The following two chapters introduce different superconductive logic families, including the logic gates, interconnect, and bias current distribution. Although particular emphasis is placed on rapid single flux quantum (RSFQ) circuits, other types of SFQ logic families and superconductive circuits are also briefly reviewed. The following three chapters discuss synchronization, fabrication, and electronic design automation (EDA) methodologies, reviewing both widely established concepts and techniques as well as recent novel approaches. The following eight chapters discuss issues, methodologies, and solutions to enable the large scale integration of complex SFQ integrated circuits and systems. Issues related to memory, synchronization, bias networks, and testability are described, and models, circuits, algorithms, and design methodologies are discussed and placed in context.

Single-flux quantum circuits are capable of transforming large scale computing systems—an increasingly important application due to the movement of data storage and processing onto remote cloud servers. Many challenges currently faced by superconductive electronics will be resolved by the increasing focus on this exciting technology, as clearly demonstrated by the large number of publications in this burgeoning area. This book aims to introduce and provide engineering intuition into the physics, devices, circuits, methodologies, and architectures to enable the development of next generation, large scale single-flux quantum systems.

Rochester, NY, USA   Gleb Krylov

Rochester, NY, USA   Eby G. Friedman

Acknowledgments

The authors would like to thank Charles Glaser from Springer for his support and encouragement. The authors express their appreciation to Dr. Mark I. Heiligman and Dr. D. Scott Holmes for their programmatic support as part of the IARPA SuperTools program. The authors would also like to deeply thank Dr. Stephen Whiteley for his valuable insight, comments, and corrections, and Jamil Kawa for his intellectual and supportive guidance over many years.

The original research work presented in this book was made possible by the Department of Defense (DoD) Agency—Intelligence Advanced Research Projects Activity (IARPA) through the U.S. Army Research Office under Contract Nos. W911NF-14-C0089 and W911NF-17-9-0001, by the National Science Foundation under Grant Nos. CCF-1329374, CCF-1526466, and CCF-1716091, AIM Photonics under Award No. 059447-007, the Intel Collaborative Research Institute for Computational Intelligence (ICRI-CI), Singapore Ministry of Education under Grant Nos. MOE2014-T2-2-105 and MOE2019-T2-2-075, and by grants from Cisco Systems, OeC, Qualcomm, Synopsys, and Google. The content of the information does not necessarily reflect the position or the policy of the U.S. Government, and no official endorsement should be inferred.

Contents

Chapter 1
Introduction

A computer is a complex system of interlinked devices. The history of computing, starting with the ancient mechanical calculators, is a history of improving both the organization of these systems and the underlying device technology. From the earliest mechanical calculators, to sophisticated mechanical systems developed during the early twentieth century, to electromechanical computers of the 1940s, to vacuum tube electronics of the 1950s, to transistors of the 1960s, to integrated circuits of the 1970s, culminating in very large scale integration (VLSI) over the past several decades—the computing industry has experienced over the last century multiple dramatic leaps in computing power. All of these leaps occurred when the switching devices at the core of the computing system evolved, followed by a long and incremental process of component and system-level enhancements. The primary incentives for enhancing modern computing systems are to further improve performance and power efficiency.

Many different materials, devices, and technologies to build a computer exist. This book is focused on one group of technologies—superconductive electronics. This technology has the potential to vastly improve the performance and efficiency of next generation computing systems. Superconductivity is a unique phenomenon that exists in specific materials at low temperatures, first discovered in 1911 by Heike Kamerlingh Onnes [1]. The electrical resistance in a superconductive material disappears below a specific temperature. Semiconductor materials, in which the resistance drastically changes due to a variety of factors, have been known since the nineteenth century [2]. Theoretical explanations for both of these unusual materials were first proposed in the 1930s with the advancement of quantum theory.

Both semiconductor and superconductor electronics have evolved in parallel. The first semiconductor switching device—the transistor—was invented in 1947 by J. Bardeen, W. Brattain, and W. Shockley [3], although proposed earlier by J. E. Lilienfeld in 1925 [4]. The first superconductive switching device— the cryotron—was developed around 1954 by D. A. Buck [5], starting the field of superconductive electronics (SCE). At that time, computers primarily utilized

© Springer Nature Switzerland AG 2022   1
G. Krylov, E. G. Friedman, *Single Flux Quantum Integrated Circuit Design*,
https://doi.org/10.1007/978-3-030-76885-0_1

vacuum tube elements. These early electronic computing machines were created for military applications, were expensive, and occupied significant area. Both of these new switching technologies, semiconductor-based and superconductor-based, could enable smaller and cheaper computers, as shown in Fig. 1.1. Although the first cryotrons, such as the device depicted in Fig. 1.1, were composed of individual wires, the simple structure of the device enabled thin film fabrication [6] and further miniaturization. The possibility of more compact and robust computers intrigued the nascent computing industry.

Integrated circuits (IC) promised to further shrink the size of switching devices. Jack Kilby in 1958 was the first to develop a semiconductor IC, where multiple transistors were integrated on a piece of germanium [8]. Around the same time, 1957 to 1962, the first superconductive ICs integrating multiple cryotrons and the first cryotron-based, relatively compact memory arrays were produced [9]. Cryotrons were based on a simple structure—two different metals separated by an insulator—as opposed to transistors, which required more precise control of the interaction between the dopant and semiconducting material. The superconductive nature of the devices meant that the current, representing a specific state, could circulate indefinitely without dissipation—an important advantage over transistor-based memory, which requires special latching circuits or regeneration (refresh) procedures [6]. Despite these major advantages of superconductivity, by the end

Fig. 1.1 *New York Times* newspaper article from February 6, 1957 [7], foreshadowing cubic foot sized computers. Ubiquitous at the time, the vacuum tube (top) is shown next to an early transistor (left) and cryotron (right)

of the 1960s, early semiconductor fabrication issues were resolved, and investments primarily concentrated on semiconductor electronics [9].

With the advancement of transistor-based electronics and integrated circuits, computing machines became more widespread and suitable for a broader variety of tasks, in particular, science and business. The number of these large (mainframe) machines, however, remained limited, and an organization often only owned a single system. Companies and universities gradually began to use these systems in a time-sharing fashion, where the computing time was divided among multiple users. With the increasing number of users, operating systems were designed to automate the time-sharing process [10]. Specialized terminals were used to input commands and programs, as well as remotely analyze the results of computations. While early terminals were essentially an input device (a teletype), more advanced terminal computers contained display and additional circuits for specialized processing [11].

With further increases in the complexity of semiconductor ICs, relatively small microcomputers (personal computers) became feasible for some of the less complex computing tasks. These machines were significantly less expensive and could be installed at a workplace or even at home. For more complex tasks, these machines were used as terminals to connect over a network to a mainframe or server computers within an organization. Another application of these smaller computing systems was embedded and on-board computers. These machines were initially primarily used for military and space applications, but later adapted to a wider variety of tasks.

These applications were enabled by the shrinking size and reduced power consumption and cost of semiconductor-based ICs. The scaling of semiconductor ICs began [12] and is continuing to this day. Cryotron electronics, requiring a cryogenic environment to operate, were not competitive for these applications, and development stalled.

A major event for superconductive electronics occurred in 1962 when Brian Josephson discovered what became known as the Josephson effect, used in a novel device—the Josephson junction (JJ) [13]. This device combined a simple structure (an insulator between two layers of metal) with sub-nanosecond switching times [14]. During the 1960s and 1970s, this switching speed was rarely achieved by contemporary bipolar, NMOS, and PMOS semiconductor digital circuits [15]. Although certain semiconductor-based switching elements, such as tunnel diodes [16], have demonstrated comparable switching speeds [17], the development of diode logic was stalled by the advances of CMOS integrated circuits. A significant research effort was concentrated at IBM during this time, and later in other companies, to develop high performance JJ-based systems [18]. This effort culminated in the late 1980s with the development of ETL-JC1 in Japan, shown in Fig. 1.2—a multi-chip four-bit JJ-based computer [19].

During the 1980s to 1990s, the continuing advancement of complementary metal oxide semiconductor (CMOS) electronics, fueled by interest from the rapidly growing computing industry, enabled mobile (portable) computers. The rapid growth of this application area, and therefore increasing investment, did not help the development of superconductive electronics. Despite the comparable size of

Fig. 1.2 Photograph of the
ETL-JC1 computer [20].
Several superconductive ICs
form a multi-chip module

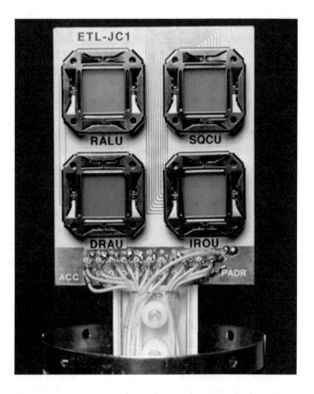

the basic devices, superconductive circuits required bulky and costly cryogenic
equipment to operate, restricting this technology to large stationary systems.
Simultaneously, semiconductor transistors, first bipolar, followed by NMOS and
then CMOS, achieved switching speeds comparable to the early superconductive
circuits. Superconductive electronics lost attractiveness due to the limitations of the
primary application area, stationary computing. The development of superconductive electronics as a technology once again slowed.

In 1986, J. G. Bednorz and K. A. Müller discovered superconductivity at a
temperature much higher than previously observed—a phenomenon called high
temperature superconductivity (HTS) [21], as opposed to conventional low temperature superconductivity (LTS). These novel ceramic materials exhibited unusual
properties which could not be explained by existing theories of superconductivity.
Significant research efforts were soon dedicated to this novel area. One aspect of
these efforts focused on the search for even higher temperature superconductors,
with an expectation that eventually room temperature superconductivity will be
discovered. Other efforts targeted HTS electronics. SCE using HTS materials
could operate at significantly higher temperatures, drastically reducing the cost of
cryogenic refrigeration. These materials and circuits, however, exhibit a number of
issues that complicate the design and fabrication of HTS circuits as compared to
LTS circuits, making scaling complexity difficult [22].

By the end of the 1980s, novel types of superconductive circuits, different from the early cryotron and JJ-based circuits, were proposed [6]. These new circuits— rapid single flux quantum (RSFQ) [23] and quantum flux parametron (QFP) [24]—exploited two primary advantages of superconductive circuits, respectively, extremely high speed and excessively low power per operation, each technology targeting high performance digital computing. Despite these advantages, the increasing speed and decreasing power consumption of CMOS circuits without requiring refrigeration, combined with the proliferation of mobile computing, prevented the widespread adoption of superconductive circuits. In addition, the novel field of HTS electronics attracted significant attention, further redirecting investment from conventional LTS circuits [6, 18]. Only a few specific niche applications, such as voltage standards and analog-to-digital conversion, remained commercially viable. Superconductive circuits for digital applications, however, continued to evolve; modern superconductive circuits have the potential to provide a significant performance advantage over conventional room temperature semiconductor-based electronics.

Computer technology also evolved as have novel ways to use computers. As the area of applications for computers widened, the target power consumption of computing systems also widened, as shown in Fig. 1.3. The rapid increase in sophistication and decrease in the size and power requirements of semiconductor ICs and systems culminated in handheld mobile computers. Combined with the communications capability of the mobile phone, also enabled by increasingly complex ICs, smart phones are currently a primary market for integrated circuits. Conversely, over the years, large scale computing systems became even larger. Modern supercomputers and data centers frequently occupy an entire building and

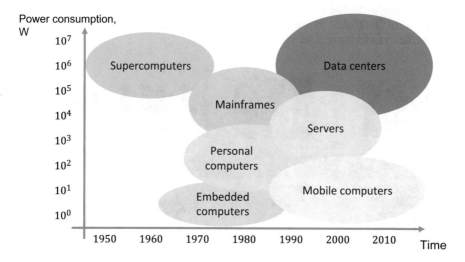

Fig. 1.3 Computing systems classified by total power consumption and approximate time of development. All of these computing applications are in common use today

consume megawatts of power [25]. From mainframes to servers to workstations to laptops to cell phones, these computing systems of all sizes and levels of power consumption are now ubiquitous, commonly used in dramatically different ways. These different computing systems are classified in Sect. 1.1 based on the total power consumption.

The continuing increase in the complexity of integrated circuits has enabled the advancement of computer networking technology and the Internet. The capability to access remote computers has been evolving for a long time, as shown in Fig. 1.4. Modern ICs support high speed wired networks and wireless communications by providing computing power for efficient encoding and transmission of data. By combining these communications technologies with highly efficient computing hardware, modern smart phones possess the capability to access remote servers to perform complex computations which would otherwise be infeasible for these small devices. These remote stationary large scale systems, so-called "cloud" systems, are an essential part of the modern mobile computing infrastructure. Users, ranging from commercial social networks to governments, depend upon this capability. Huge amounts of data are now offloaded to these distant server farms. As described in Sect. 1.2, cloud computing has now become a dominant and important stationary application where high speed and, importantly, low energy are of seminal currency.

As the scaling of CMOS systems becomes progressively more expensive, multiple technologies have been proposed to continue to improve performance and power efficiency. Low temperature and cryogenic electronics specifically target large scale stationary systems and offset the cost of the cryogenic refrigeration with higher power efficiency. Modern superconductive electronics promise at least two orders of magnitude improvement in energy efficiency as compared to conventional semiconductor-based supercomputers [26]. Combined with other benefits, as described in Sect. 1.3, these circuits are potentially a highly attractive solution for future data centers and supercomputers and are the focus of this book. An outline of this book is provided in Sect. 1.4.

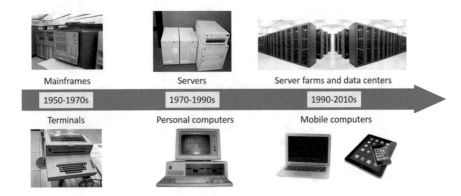

Fig. 1.4 Evolution of remote access in computing systems

1.1   Types of Computing Systems

The diversity of applications for computing machines has led to the parallel development of specialized computing systems. In parallel, the complexity of integrated
circuit technology has greatly increased over time. Most modern processors contain
millions to many billions of transistors and therefore exhibit VLSI complexity.
These high complexity processors support different functional purposes and/or
power requirements. All modern computing systems, regardless of complexity, can
be divided into several groups based on the power requirements. Each group of
computing systems, however, possesses certain design objectives and limitations. In
this section, these different groups and related limitations are reviewed.

Small mobile devices (e.g., handheld computers, phones, embedded sensors,
laptops), shown at the bottom of Fig. 1.3, are typically battery operated. A design
emphasis of these devices is on providing sufficient time before the battery is
discharged. Another design objective is small size. Since the battery capacity
increases the size of the device, power consumption in mobile devices is severely
limited. Due to this issue of size, these devices also lack an efficient cooling system,
limiting the ability to remove heat. The integrated circuits used in these systems
are limited by the thermal design power (TDP)—the maximum amount of heat
dissipated by an IC. The computing power in these systems is often secondary to
these objectives.

Medium-sized computers (e.g., personal desktop, large embedded computers,
and small servers), occupying the middle of Fig. 1.3, are typically directly connected
to a power grid. This feature relaxes the limitation on power consumption. The
power requirements of these systems, although still a concern, translate to the cost
of electricity and on-chip current densities and heat dissipation (and therefore the
cooling system). The size requirements of these stationary computers are also less
significant. These computers typically require a dedicated active cooling system,
enabling the use of high TDP ICs. The primary factors limiting the performance
of these computers are the cost of the hardware, electricity, and physical size. Due
to these limitations, the number of processing units (individual ICs and/or on-chip
cores) in these systems tends to be small.

Large scale computing systems (e.g., large servers, mainframes, supercomputers,
data centers), depicted at the top of Fig. 1.3, are directly connected to the power grid
and are typically not limited by physical size, only the cost and total power consumption. These machines are primarily used for military, business, and scientific
applications and are often highly specialized for a particular task. The performance
of these computers scales with the number of ICs, where the performance is limited
by the cost of the hardware. Large scale computers employ advanced networking
protocols to connect the processing units, as well as sophisticated cooling systems.
The primary metrics for these systems are performance and power efficiency,
typically measured in operations per watt.

Large scale computers exhibit the highest power efficiency due to the fixed costs
associated with heat dissipation and removal. Modern large scale systems frequently

exploit this efficiency to support many thousands to millions of simultaneous independent users in a cost-effective manner, as discussed in Sect. 1.2. Due to fewer design limitations, these systems can utilize novel exploratory architectures and emerging technologies to further increase performance, as discussed in Sect. 1.3. The primary focus of this book is the application of superconductive electronics to improve the efficiency of these large scale (high total power) computing systems.

1.2   Remote (Cloud) Computing

A significant recent trend in the computing industry is the increasing dependence of small and medium scale computers on large scale distant computing systems. While some of these smaller systems were previously used to connect to mainframes and supercomputers, as shown in Fig. 1.4, these remote connections were frequently limited by distance and bandwidth. The majority of personal computers operated relatively independently from other computers. With continuing advances in communications technologies and the Internet, remote servers now appear to be almost seamlessly accessible worldwide. Data storage and processing tasks are increasingly being offloaded onto these remote (cloud) computing systems. Handheld and other mobile devices are dependent on and function as terminals to access this remote computing power.

The primary reason for the partial shift of the computation process to remote computers is the aforementioned limitations in performance and storage of small and medium scale systems. Small devices are incapable of storing and processing as much data as large systems. The scaling of CMOS transistors and therefore the increase in IC performance has slowed due to physical limitations of the transistor itself and related connections [27]. The large power (and therefore heat) dissipated by many billions of transistors limits the performance of modern processors. Significant portions of modern ICs are temporarily disabled to avoid overheating [28]. This issue is known as "dark silicon," where only a portion of an IC can operate at any time. The energy efficient systems-on-chip used in small scale computers are highly sensitive to this effect since the cooling systems of these ICs are less efficient. This effect is expected to continue to slow the scaling of CMOS technologies [27].

Large scale computers are less susceptible to this problem. Due to the large physical size, these systems employ highly efficient cooling methods. Different heat-generating components can be placed far from each other, reducing the power density within these systems. Modern data centers and supercomputers are also frequently located in advantageous areas—places with a cold climate, available water for cooling, and less expensive and/or highly available electricity. Modern software programs and applications frequently employ remote (cloud) functionality. If the current trend, as depicted in Fig. 1.4, continues, the number and significance of these remote computing systems will further increase. Large scale, high power computing systems are therefore expected to greatly increase in importance.

As previously mentioned, the primary metrics for these large scale systems are performance and power efficiency. For commercial applications, power efficiency is paramount, as a higher efficiency reduces the maintenance cost of these large scale computing systems. Military and scientific applications also benefit from power efficiency, as the total power consumed by modern supercomputers is on the order of tens of megawatts [25]. The slowdown of CMOS scaling also affects these high power systems, and multiple beyond-CMOS technologies are currently under development [29]. These alternative technologies are seen as either a replacement or enhancement of conventional CMOS-based electronics, often targeting specific applications. This book is focused on one family of these technologies—superconductive electronics.

1.3 Superconductive Electronics

Multiple technologies are currently being considered to supplement conventional CMOS circuits targeting specific applications. These technologies range from near mature magnetic and spintronic technologies to highly exploratory and unusual approaches to computation, such as biological systems. Low temperature (cryogenic) electronics in general is one of these promising technologies. Both semiconductor and superconductor circuits inherently dissipate much lower power when operating at cryogenic temperatures due to the reduced resistance of the metals, higher carrier mobility in the transistors, and the ability to operate at lower supply voltages.

These considerable advantages of cryogenic operation are not commonly used in modern room temperature computers. The primary issue is the requirement to maintain a low temperature environment. Circuits designed for low temperature operation are either cooled by a closed cycle refrigerator [30] or submerged into cryogenic liquids [31]. Closed cycle refrigerators require significant power to operate, increasing the cost of the overall system. Systems utilizing liquid cooling require a continuous supply of cryogenic liquids, also increasing cost; nitrogen (liquid at 77 K) or helium (liquid at 4 K) is typically used, depending upon the target temperature. These cooling costs are reduced in cold environments, such as outer space, which is an important and early application area of low temperature electronics [32].

Superconductive circuits exhibit numerous benefits as compared to both conventional and low temperature CMOS circuits. Superconductive circuits require extremely low energy per operation [33], as shown in Fig. 1.5. Superconductivity also produces lossless interconnect and dissipates zero static power. In addition to energy efficiency, superconductive electronics provide numerous additional benefits. Certain types of superconductive circuits can operate at clock frequencies exceeding 100 Ghz [34]. Apart from digital computation, this high speed switching is also applicable to signal processing applications [35]. Certain types of superconductive circuit families can operate in an adiabatic reversible manner, enabling

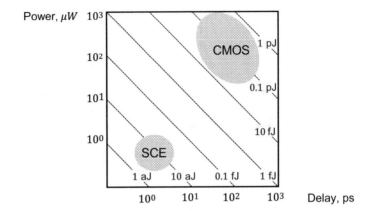

Fig. 1.5 Delay (performance) and power (energy dissipation) per operation for CMOS and SCE technologies [40, 41]

extremely high energy efficiency [36]. Certain superconductive circuits can detect extremely low magnetic fields [37] or single photons [38]. SCE also forms a natural interface with superconductive quantum circuits, behaving as an interface between conventional room temperature electronics and millikelvin superconductive qubits [39].

The primary obstacle to the wide adoption of SCE is the low temperature. The majority of modern superconductive circuits are based on niobium, typically cooled to approximately 4 K [33]. The power required to maintain this extremely low temperature is significant, and the necessary equipment also requires significant space. These properties limit the application of superconductive circuits to large scale systems, where the size and total cost are secondary to energy efficiency and cost per operation. After considering the energy required for refrigeration, superconductive circuits exhibit two to three orders of magnitude lower energy per operation than state-of-the-art conventional supercomputers [25, 35].

A large scale system (such as a supercomputer) based on superconductive electronics however poses a significant engineering challenge. Modern supercomputers typically contain thousands of separate processing units [42]. A similar architecture would be necessary for an SCE-based system. It is feasible to place these components into a shared cryogenic environment to reduce the length of the interconnect and the amount of cryogenic equipment. As the cost of extracting heat from a system, however, increases at lower temperatures, these systems require multiple layered environments with progressively lower temperatures [33]. A temperature layered system is schematically shown in Fig. 1.6. Only relevant components are placed within each temperature level. For example, although superconductive qubits require a temperature on the order of a few millikelvins to operate [43], classical information extracted from a quantum circuit can be further processed at a higher temperature [39]. In this way, only a small portion of the system needs to be cooled to millikelvin temperatures.

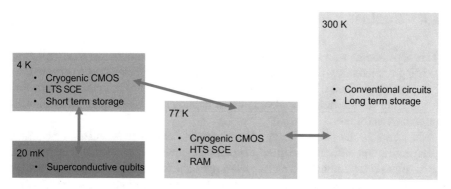

Fig. 1.6   Large scale, layered cryogenic computing system [33, 39]

Significant time, effort, and resources have been dedicated to developing CMOS electronics over the past 60 years. Advances in electronic design automation (EDA) (or computer-aided design (CAD)) algorithms, tools, and methodologies have supported the increasing complexity of CMOS circuits. These advances have led to the huge number and variety of applications, attracting additional investment in a positive feedback loop.

Superconductive circuits have not experienced similar development. With the small number of organizations and research groups working on SCE, superconductive circuits are currently significantly less mature as compared to conventional CMOS circuits. Unlike some other novel beyond-CMOS technologies, SCE is not recently emergent and has been studied for many years [18]. The physical properties of superconductive materials and devices are now well understood; considerable engineering effort is, however, required to make large scale SCE systems practical and cost effective.

SCE is capable of transforming large scale computing systems—an increasingly important application due to the aforementioned shift of data storage and processing onto remote servers. The time is now for superconductive technology to fill this niche for a stationary, high speed, and extremely low energy computing technology. Add to this wonderful juxtaposition of evolving technologies and applications the combination of classic space-based satellites and the exciting field of quantum computing requiring millikelvin circuit interfaces (both applications where cryogenic operation is either naturally available or necessary), it has become clear that superconductive electronics is at the cusp of becoming a mainstream technology. What was needed is a high volume application and a technology ready to be matured. Now both are available.

1.4 Outline

This book begins with several chapters introducing superconductive electronics and existing related design techniques. Multiple circuits and techniques to enable high complexity circuits are described in the approximate order of increasing abstraction level, from device models and circuits to architectural techniques. This book is organized as follows.

In Chap. 2, the phenomenon of superconductivity is introduced. A historical perspective and different theories of superconductivity are presented. The physical properties of superconductive materials important for electronics applications are discussed. As compared to conventional transistor-based circuits, superconductive electronics utilize a different set of basic devices as building blocks of larger circuits. These basic devices are introduced in this chapter.

Superconductive circuits are summarized in Chap. 3. Superconductive quantum interference devices (SQUIDs)—a basic circuit ubiquitous in superconductive electronics—are described. Many distinct logic families of superconductive digital logic exist; each family utilizes a drastically different signaling convention, basic gates, and synchronization and power distribution methods, and target different application areas. These logic families are reviewed in Chap. 3. Different types of storage elements and memory cells suitable for use in a cryogenic environment are also discussed.

In Chap. 4, the most popular type of superconductive digital logic—rapid single flux quantum (RSFQ) logic—is described. Transmission lines for single flux quantum (SFQ) pulses are also introduced. The operation of basic logic gates and flip flops in the RSFQ logic family is described. The distribution of bias currents, necessary to operate these circuits, is also discussed along with energy efficient bias distribution techniques.

In Chap. 5, synchronization of superconductive circuits is reviewed. Existing pulse-based clock distribution topologies are introduced and related tradeoffs are described. Asynchronous techniques which do not require a global clock are also discussed. Synchronization in AC-biased circuits, utilizing a different clocking mechanism, is also described.

In Chap. 6, manufacturing of superconductive circuits is reviewed. Different steps and materials used in this process are described. Challenges unique to superconductive electronics are highlighted. Important features of modern superconductive fabrication technologies are discussed and compared to the fabrication of semiconductor integrated circuits.

In Chap. 7, EDA methodologies, techniques, and algorithms used in superconductive electronics are discussed [44]. The semi-custom standard cell-based design flow, common in conventional CMOS circuits, is increasingly widely adopted in modern superconductive circuits. Differences and issues in computer-aided design flows as compared to CMOS design methodologies are highlighted in Chap. 7. The most common stages of these design flows, from high level simulation to physical layout, are described. These stages are grouped into three areas—

simulation/modeling, synthesis, and verification. Modern approaches and tools for superconductive circuits are reviewed for each of these areas, including both manual and automated techniques.

The superconductor-ferromagnetic transistor (SFT) is a promising device capable of interfacing with superconductive memory arrays. To build complex circuits utilizing an SFT, an efficient and compact circuit-level model is necessary. In Chap. 8, a compact model for the SFT device is described and compared to experimental data [45]. The transient behavior of this model is also compared to the expected behavior.

Modern superconductive circuits utilize more than ten superconductive layers for gates and interconnect. Many sources of inductive coupling noise exist within this environment [46]. In Chap. 9, these sources are characterized, and the effects of inductive coupling noise on different circuit structures are described. Guidelines to mitigate the deleterious effects of noise coupling are presented.

Compact and efficient memory is an important issue in superconductive electronics. Spin-based memory can be used in a cryogenic environment, combined with an appropriate superconductive driver device. In Chap. 10, a sense amplifier for reading a cryogenic spin-based memory cell is described [47]. The proposed sense amplifier exploits the specific shape of the readout waveform within a well known and compact superconductive A/D converter topology.

Dynamic single flux quantum (DSFQ) is a novel approach for asynchronous SFQ-based logic, capable of drastically reducing the complexity of clock networks as compared to conventional synchronous RSFQ circuits. DSFQ logic is introduced and a DSFQ-based majority gate is described in Chap. 11. A compact DSFQ majority gate is proposed to reduce area and increase the performance of DSFQ circuits [48].

DSFQ logic is a promising asynchronous type of SFQ logic. Logic synthesis of DSFQ circuits significantly differs from both CMOS and conventional RSFQ circuits. In Chap. 12, enhancements to the automated synthesis of DSFQ circuits are described. The concept of path balancing, widely used in RSFQ circuits, is explored for DSFQ logic. An area efficient wave pipelining methodology is also proposed in Chap. 12 to increase system throughput.

Bias distribution is essential for energy efficient and high performance superconductive circuits. Several energy efficient bias distribution techniques have been proposed for RSFQ circuits. In Chap. 13, the operation of energy efficient RSFQ bias networks is described, parametric trends are discussed, and a set of design guidelines for these networks is introduced [49, 50]. These guidelines, along with a distributed approach to bias networks, are intended to automate the synthesis of these structures.

Large bias currents is a significant obstacle to the large scale integration of DC-biased superconductive circuits. In Chap. 14, current recycling (or serial biasing) is introduced. A methodology for automated partitioning superconductive integrated circuits during the placement process is described [51]. Two different approaches to partitioning are evaluated and compared.

Synchronization is an important issue in large scale multi-gigahertz circuits. Compartmentalization of different components within a system is a common engineering approach to manage complexity. Globally asynchronous, locally synchronous (GALS) clocking techniques isolate the complexity of synchronous clock networks to individual, smaller blocks, thereby simplifying the design process [52]. A GALS clocking scheme for RSFQ circuits, along with shared interconnect for large scale systems-on-chip, is described in Chap. 15.

Different steps of the design flow as well as variations in the fabrication process can introduce errors and defects into an electronic circuit. A large scale system invariably contains many defects. The ability to detect these errors and determine the cause is important to produce reliable systems. In Chap. 16, the design-for-testability (DFT) techniques are introduced [53, 54]. These features are widely used in CMOS circuits to detect errors after fabrication and determine the scope of correct circuit operation. Similar DFT features suitable for superconductive circuits are proposed in Chap. 16.

Chapter 2
Physics and Devices of Superconductive Electronics

2.1 Introduction

The phenomenon of superconductivity—a material exhibiting zero electrical resistance—was discovered in 1911 by Heike Kamerlingh Onnes [1]. Onnes was the first person to liquefy helium. During his experiments with mercury wires at liquid helium temperatures (4.2 K), the resistance of these wires completely vanished. For this discovery, Onnes was awarded a Nobel Prize in 1913.

Multiple theories and explanations have been proposed to explain this phenomenon and provide predictive expressions for the unusual effects occurring in superconductors. Based on these theories, different superconductive electronic devices have been developed. In this chapter, these theories, effects, and devices are discussed.

This chapter is organized as follows. In Sect. 2.2, many of the existing theories of superconductivity are described in chronological order. In Sect. 2.3, the unusual properties of superconductive materials are reviewed with an emphasis on those effects relevant to superconductive electronics. In Sect. 2.4, the primary device used in superconductive digital circuits—the Josephson junction—is introduced, and the properties and operation of this device are discussed. In Sect. 2.5, other cryogenic devices commonly used in modern superconductive circuits and systems are described along with some typical applications. A brief summary of this chapter is provided in Sect. 2.6.

2.2 Theories of Superconductivity

In this section, three theoretical frameworks describing conventional superconductivity are reviewed in chronological order. Among these theories are the London theory, Ginzburg-Landau theory, and Bardeen-Cooper-Schrieffer theory.

© Springer Nature Switzerland AG 2022
G. Krylov, E. G. Friedman, *Single Flux Quantum Integrated Circuit Design*,
https://doi.org/10.1007/978-3-030-76885-0_2

Some important parameters and effects described by these theories are discussed in Sect. 2.3.

2.2.1 London Theory

The London theory, named after brothers Fritz and Heinz London, who first described this theory in 1935, is the first and one of the simplest phenomenological explanations of superconductivity [55]. In this section, the London equations are derived and discussed, and some of the primary effects characterizing superconductors based on these expressions are described.

2.2.1.1 Derivation of London Equations

Consider a derivation of Ohm's law for a normal metal. From Newton's law,

$$m\frac{dv}{dt} = eE - \frac{mv}{\tau}, \tag{2.1}$$

where m, v, and e are, respectively, the mass, velocity, and charge of the carrier. The first term on the right-hand side is the force exerted by the electric field, and the second term is the scattering with mean time τ. For a steady state condition, the electron drift velocity is constant; therefore,

$$eE - \frac{mv}{\tau} = 0, \tag{2.2}$$

$$v = \frac{eE\tau}{m}. \tag{2.3}$$

The current density for n carriers is

$$J = nev = \frac{ne^2\tau}{m}E = \sigma E, \tag{2.4}$$

where $\sigma = \frac{ne^2\tau}{m}$ is the conductivity. Expression (2.4) is the standard Ohm's law for a normal metal.

If the electrons in a superconductor do not scatter, and the second term on the right-hand side of (2.1) is equal to zero,

$$\frac{dv}{dt} = \frac{eE}{m}. \tag{2.5}$$

The time derivative of the current density is

$$\frac{\mathrm{d}J}{\mathrm{d}t} = n_s e \frac{\mathrm{d}v}{\mathrm{d}t} = \frac{n_s e^2}{m} E. \tag{2.6}$$

Expression (2.6) is the first London equation, and n_s is the phenomenological density of superconductive electrons.

To obtain the second London equation, apply a curl operation to the first London equation,

$$\nabla \times \frac{\mathrm{d}J}{\mathrm{d}t} = \frac{n_s e^2}{m} \nabla \times E. \tag{2.7}$$

From Maxwell's equations,

$$\nabla \times E = -\frac{\mathrm{d}B}{\mathrm{d}t}. \tag{2.8}$$

Substituting (2.8) into (2.7) and integrating both sides with respect to time,

$$\nabla \times \frac{\mathrm{d}J}{\mathrm{d}t} = -\frac{n_s e^2}{m} \frac{\mathrm{d}B}{\mathrm{d}t}, \tag{2.9}$$

$$\nabla \times J = -\frac{n_s e^2}{m} B. \tag{2.10}$$

Expression (2.10) is the second London equation.

2.2.2 Ginzburg-Landau Theory

The Ginzburg-Landau (GL) theory, developed by Vitaly Ginzburg and Lev Landau in 1950, is a more complex phenomenological theory describing the behavior of superconductors [56]. In the London theory, the density of the superconductive electrons n_s is not derived and is treated as a phenomenological parameter. The dependence of this density on temperature and field can be described based on GL theory.

GL theory is an application of the more general Landau theory of phase transitions. In the general Landau theory, the phase transitions are described using a parameter distinguishing an ordered state from a disordered state—an order parameter. For example, magnetization is the order parameter in ferromagnetic phase transitions. With this parameter as well as other thermodynamic variables, the corresponding thermodynamic equations can be derived.

In general, superconductivity is a second-order phase transition, where an ordered state emerges from a disordered state at lower temperatures [57]. Superconductivity is similar to a different cryogenic phenomenon—superfluidity, allowing superconductive electrons to be described as a charged superfluid [58].

2.2.3 Bardeen-Cooper-Schrieffer (BCS) Theory

BCS theory, developed in 1957 by John Bardeen, Leon Cooper, and John Robert Schrieffer, is the first theory explaining the origin and microscopic properties of superconductivity [59]. In this theory, superconductivity is described as the condensation of bound pairs of electrons, where the pairs are connected through electron-phonon interactions. This theory won the Nobel Prize in Physics in 1972.

In simple terms, electrons in superconductive materials form bound pairs, called Cooper pairs [60]. One electron traveling through a material distorts the lattice of this material. This distortion produces a localized positive charge density. This positive charge density in turn attracts another electron. This interaction between electrons and the lattice (phonons) binds electrons into Cooper pairs.

The physical size of a Cooper pair, the distance between the electrons, is described by the superconductive coherence length ξ. The distance between the paired electrons is relatively large, on the order of tens to hundreds of nanometers. Many pairs existing in a material overlap, producing a collective condensate exhibiting the same quantum state [61].

2.3 Properties of Superconductive Materials

In this section, unusual properties of superconductive materials are discussed. In Sects. 2.3.1 to 2.3.4, the unique properties of superconductors—the Meissner effect, quantization of magnetic flux, and existence of an energy gap and quasiparticle currents—are discussed. In Sects. 2.3.5 to 2.3.7, important parameters characterizing superconductive materials—the London penetration depth, coherence length, and critical current—are reviewed. In Sects. 2.3.8 and 2.3.9, different classifications of superconductive materials are introduced. In Sect. 2.3.10, the kinetic inductance—an important inductance component in superconductive circuits—is discussed.

2.3.1 Meissner Effect

The Meissner effect, also known as the Meissner-Ochsenfeld effect, is the expulsion of magnetic field from a bulk superconductor [62]. Superconductors expel a magnetic field when the temperature is below a critical temperature. This phenomenon was experimentally discovered in 1933 by Walther Meissner and Robert Ochsenfeld [63] and can be derived from the London equations.

Consider Ampere's circuital law,

$$\nabla \times B = \mu_0 J. \tag{2.11}$$

Applying a curl to both sides of the expression,

$$\nabla \times \nabla \times B = \mu_0 \nabla \times J. \tag{2.12}$$

$$\nabla(\nabla \cdot B) - \Delta B = \mu_0 \nabla \times J. \tag{2.13}$$

Substituting the second London equation, (2.10), into (2.13),

$$\nabla(\nabla \cdot B) - \Delta B = \mu_0 \frac{n_s e^2}{m} B. \tag{2.14}$$

From Maxwell's law, $\nabla \cdot B = 0$; therefore,

$$\Delta B = \frac{\mu_0 n_s e^2 B}{m} = \frac{B}{\lambda_l^2}, \tag{2.15}$$

where

$$\lambda_l = \sqrt{\frac{m}{\mu_0 n_s e^2}} \tag{2.16}$$

is the London penetration depth described in Sect. 2.3.5. The solution for the differential equation (2.15) is

$$B = B_0 e^{\frac{-x}{\lambda_l}}, \tag{2.17}$$

where x is the depth within the superconductor and B_0 is the surface field. Expression (2.17) shows that the magnetic field within a superconductor decreases exponentially with depth.

A magnetic field is expelled from a superconductor by creating currents along the surface of the material. This effect is schematically illustrated in Fig. 2.1. These currents oppose any flux penetration into the bulk material. These currents do not dissipate over time due to the zero resistance of the material and are therefore called persistent currents.

The Meissner effect emphasizes the difference between an ideal perfect conductor, as a superconductor was imagined at that time, and a superconductor. In a perfect conductor, the magnetic fields that exist at the time of transition into the

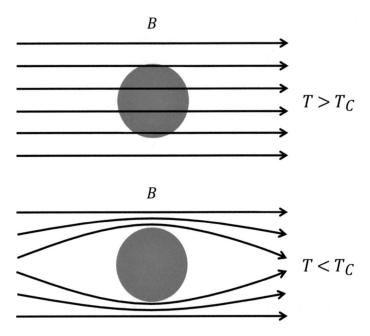

Fig. 2.1 Meissner effect. Magnetic field lines are expelled from the superconductive material [64]

superconductive state become constant within the material, resisting any change. In a superconductor, any applied flux is expelled during a transition.

2.3.2 Quantization of Magnetic Flux

Fritz London was the first person to propose that superconductivity is a quantum effect. While the London equations are phenomenological, the superconductive state was assumed to be similar to the ground state of an atom. The motion of all electrons in the superconductive state is correlated and is described as a single-valued wave function. From this perspective, London predicted that the magnetic field can only penetrate superconductive loops in quantized amounts—magnetic flux quanta.

Consider a wave function $\Psi = \Psi_0 e^{i\phi}$ describing these superconductive electrons. If the phase ϕ changes by 2π, no physical properties change. In a superconductive loop, the phase and magnitude of the wave function are different at different locations within the loop; the total phase change around the loop, however, can only exist in integer multiples of 2π. This phenomenon is the quantization of magnetic flux, where an integer multiple describes the number of flux quanta within a loop, as shown in Fig. 2.2.

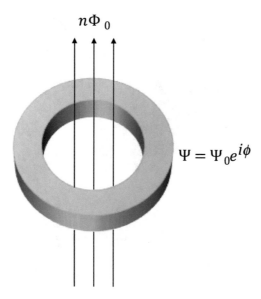

Fig. 2.2 Quantization of magnetic flux

An expression for a quantum of magnetic flux Φ_0 is

$$\Phi_0 = \frac{h}{2e} = \frac{\pi\hbar}{e},$$ (2.18)

while the numerical value of Φ_0 is

$$\Phi_0 \approx 2.0678 \times 10^{-15}\,\text{Wb} = 2.0678 \times 10^{-15}\,\text{V}\times\text{s}.$$ (2.19)

In the context of superconductive electronics, Φ_0 is often described as $2.07\,\text{mV}\times\text{ps}$.

2.3.3 Energy Gap

The collective behavior of the Cooper pairs produces an energy gap for single-electron excitations, as shown in Fig. 2.3. The energy of any single electron cannot be arbitrarily changed, since the energy of the other electrons would also need to change. This characteristic prevents the scattering of single electrons and produces dissipationless supercurrent.

The energy gap Δ is an important parameter of a superconductive material. The energy gap describes the lower energy to form a superconductive state as compared to the material remaining in a normal state. The energy gap is largest at 0 K and gradually becomes zero at the critical temperature T_c.

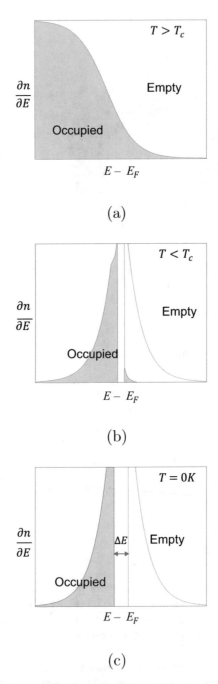

Fig. 2.3 Density of states in a superconductor for different temperatures; (**a**) $T > T_C$, (**b**) $T < T_C$, and (**c**) $T = 0$ K. The energy gap ΔE is the empty space around the Fermi level E_F

2.3.4   Quasiparticles

An ideal superconductor can only exist at zero temperature and voltage. For finite temperatures, some of the Cooper pairs are broken by thermal excitations. Voltages above 2Δ also break Cooper pairs. These broken pairs combine the individual properties of electrons and holes and are called Bogoliubov quasiparticles [65].

In superconductive electronics, these quasiparticles contribute to the electrical properties of devices. Quasiparticles scatter within a lattice; therefore, quasiparticle current is the normal current exhibiting dissipation.

2.3.5   London Penetration Depth

The Meissner effect prevents magnetic flux from penetrating into a bulk superconductor. Any magnetic fields incident on the surface of a material exponentially decay over distance. The London penetration depth describes the rate of this decay.

Consider a superconductive half plane, where $x < 0$ corresponds to an empty space and $x > 0$ corresponds to a superconductive material, as shown in Fig. 2.4. A magnetic field is applied along the z axis in the empty space. From the London and Maxwell equations,

$$B(x) = B_0 e^{(-x/\lambda_L)}, \tag{2.20}$$

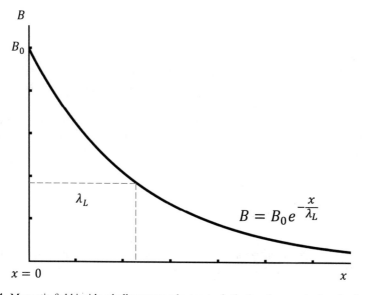

Fig. 2.4 Magnetic field inside a bulk superconductor. λ_L is the London penetration depth

$$\lambda_l = \sqrt{m/\mu_0 nq^2},\tag{2.21}$$

where B_0 is the magnetic field incident on the surface and λ_l is the London penetration depth, corresponding to the distance at which a magnetic field within a superconductor decays by e (≈ 2.72).

2.3.6 Critical Field and Critical Current

A critical current (and the corresponding critical field) is one of the primary parameters characterizing superconductive materials and devices. The critical current (magnetic field) describes the maximum current (magnetic field) that can be applied into (across) a superconductor without changing the properties of the material. The behavior of a superconductive material at fields greater than the critical magnetic field H_c depends upon the type of material, as described in Sect. 2.3.8.

The critical field depends upon the temperature. At the critical temperature, even the weakest magnetic field prevents a material from achieving a superconductive state. The maximum critical field occurs at absolute zero [66].

The critical current of a material is related to the critical field due to the Meissner effect. All currents induce magnetic fields. When the applied current induces a magnetic field with a strength exceeding the critical field, the material transitions into the normal state.

2.3.7 Coherence Length

The coherence length ξ of a superconductor provides a scale of length for superconductivity. If a material consists of both superconductive and normal regions, superconductive effects in the normal metal gradually disappear over the coherence length, as depicted in Fig. 2.5.

The coherence length was originally introduced by A. Pippard in 1950 [67] to introduce nonlocal effects into the London equations [68]. This coherence length also appears in similar but different forms in both the GL and BCS theories. The coherence length in GL theory is

$$\xi = \sqrt{\frac{\hbar^2}{2m\alpha(T)}},\tag{2.22}$$

where α is a phenomenological constant. Unlike the Pippard coherence length, the GL and BCS coherence lengths are dependent on temperature.

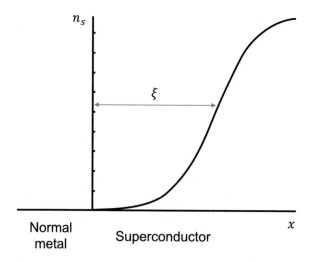

n_s

ξ

Normal
metal   Superconductor x

Fig. 2.5 Density of superconductive electrons at the boundary of a superconductive region. The coherence length is ξ

2.3.8 Type-I and Type-II Superconductors

Based on the ratio of the London penetration depth to the coherence length $\frac{\lambda_l}{\xi}$, superconductive materials can be divided into two types. This ratio also affects the behavior of a superconductor within a high magnetic field.

In type-I superconductors, $0 < \frac{\lambda_l}{\xi} < 1/\sqrt{2}$. In these materials magnetic fields greater than H_c disrupt the superconductive state. These materials are typically pure metals and exhibit a low critical temperature (< 10 K).

In type-II superconductors, $\frac{\lambda_l}{\xi} > 1/\sqrt{2}$. In these materials, two different critical fields exist—H_{c1} denotes the appearance of the mixed state, and H_{c2} denotes the field at which superconductive properties disappear. In the mixed state, a magnetic field stronger than H_{c1} penetrates into the material in the form of individual lines of magnetic flux, surrounded by a circulating supercurrent. This quantized magnetic field penetrates through non-superconductive regions, while the rest of the material remains superconductive. These circulating supercurrents are also called fluxons or Abrikosov vortices [69]. The density of these vortices in the material depends upon the external field. At fields greater than H_{c2}, the material loses any superconductive properties.

Some superconductive materials commonly used for electronic applications are listed in Table 2.1. Modern superconductive electronics are primarily based on niobium, which is a type-II material [70]. Earlier Josephson junctions were based on lead, a type-I material [71]. Superconductive qubits are typically built using aluminum, also a type-I material [72].

Table 2.1 Superconductive
materials commonly used in
electronics

Material	T_C, K	H_C, T	Type
Aluminum (Al)	1.14	0.01	I
Niobium (Nb)	9.22	0.82	II
Lead (Pb)	7.26	0.08	I
Niobium nitride (NbN)	16	~15	II

2.3.9   Low and High Temperature Superconductors

Since the discovery of superconductivity, many materials have been found to
exhibit superconductive properties. Many superconductive applications, both small
scale electronic microsystems and large scale systems, have been proposed and
are currently utilized. The major obstacle for many of these applications is the
cryogenic refrigeration required to maintain superconductive operation. Closed
cycle refrigerators require significant power, while helium refrigerators require a
constant supply of helium, and all approaches require insulated chambers [73].
Significant research efforts are therefore focused on the discovery of materials that
can operate at higher critical temperatures, with the ultimate goal to find a room
temperature superconductor.

A breakthrough in these efforts was made in 1986 by J. Bednorz and K.
Müller from IBM [21]. They discovered superconductivity in copper oxide ceramics
operating at temperatures above 35 K. Soon after this discovery, many similar
materials were found. The highest critical temperature for these materials to date
under normal pressure is 133 K [74]. These and similar materials are called high
temperature superconductors (HTS), as opposed to conventional low temperature
superconductors (LTS).

No strict definition of HTS and LTS materials exists. Those materials with a
critical temperature higher than 77 K (the boiling temperature for liquid nitrogen)
are typically called HTS, although exceptions exist. HTS materials exhibit many
properties unusual in conventional LTS materials, and as yet no complete theory
exists describing the behavior of high temperature superconductivity. All known
HTS materials are type-II superconductors.

Many applications for HTS materials exist. These materials are commonly used
in areas where extremely high magnetic fields are required—magnetic resonance
imaging (MRI), magnetic levitation, and particle accelerators. Many electronic
applications for these materials also exist. One of the most commonly used
HTS materials is yttrium barium copper oxide (YBCO). This material exhibits
superconductive properties at 92 K. The use of HTS materials for integrated digital
circuits is challenging, however, due to difficulties in fabrication and increased noise
coupling as compared to the 4 K environment.

2.3.10 Kinetic Inductance

In conventional circuits, the inductance typically refers to the storage of energy in a magnetic field produced by a current. Kinetic inductance is a different form of energy storage, produced by the motion of carriers.

In a normal conductor, the carriers are accelerated within an electric field. The carriers are simultaneously scattered by collisions with the ions within the lattice. In a steady state condition, these effects produce a constant drift velocity. If the electric field changes or disappears, electrons exhibit inertia, resisting the change—similar to a magnetic inductance. This inertia of charge carriers is kinetic inductance. The total inductance is the sum of the conventional magnetic inductance and the kinetic inductance.

Conventional materials also exhibit a kinetic inductance. In these materials, this inductance only becomes significant when the electron relaxation time is comparable to the period of change in the current. In most metals, this condition is satisfied at terahertz frequencies.

In superconductors, however, no scattering exists for Cooper pairs, and the electron relaxation time is infinite. The kinetic inductance is therefore significant and can dominate the total inductance, even at low frequencies and DC current.

2.4 Josephson Junctions

The primary device used in modern superconductive electronics is the Josephson junction (JJ). An introduction to Josephson effects, JJ dynamics, and related circuit behavior is described in this section.

2.4.1 Josephson Effects

The Josephson effect was discovered by Brian Josephson in 1962 [13], who was awarded the Nobel Prize for this discovery in 1973. The DC Josephson effect is a phenomenon caused by tunneling of Cooper pairs through an insulating barrier or a weak link (a constriction or point contact). This tunneling produces a supercurrent through a barrier without an applied voltage.

A Josephson junction (JJ) is a device that exhibits this Josephson effect. The most common type of JJ in modern superconductive circuits is a thin ($\sim$ nm) insulating film sandwiched between two superconductive contacts, as shown in Fig. 2.6. This structure is straightforward to fabricate in modern photolithographic fabrication facilities [75]. JJs can also be based on other topologies, for example, the insulating material can be replaced by a weak link, normal metal, or other materials, resulting in different device properties [76, 77].

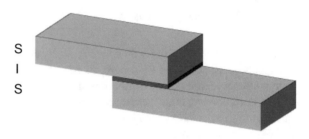

Fig. 2.6 Thin film Josephson junction. I denotes an insulating material, and S is a superconductive material

One of the major parameters characterizing a JJ is the critical current I_C. This parameter affects the maximum supercurrent conducted by a JJ. In a controlled environment, the critical current of a JJ primarily depends upon the physical area. The critical current of a JJ also exhibits a dependence on the temperature and applied magnetic field.

Most JJs in a superconductive IC exhibit similar critical currents (the same order of magnitude). Moreover, due to noise effects, these currents cannot be arbitrarily low and are typically on the order of 100 μA in modern circuits [78]. As the critical current primarily depends on area, an increase in the critical current density J_c is the primary source of scaling in superconductive circuits. The critical current density J_c depends upon the fabrication process and is a seminal parameter characterizing a fabrication process for manufacturing superconductive electronics.

An important parameter of a JJ is the phase difference across the junction (or the phase of a JJ). As all Cooper pairs in a superconductor share the same quantum mechanical state, these pairs also exhibit the same phase. The difference in phase between the wave function of the superconductive electrons at the two electrodes of a JJ affects the dynamic behavior of a JJ. The basic Josephson equations characterizing the current and voltage across a JJ are

$$I(t) = I_c \sin(\phi(t)), \tag{2.23}$$

$$U(t) = \frac{\hbar}{2e} \frac{\partial \phi}{\partial t} = \frac{\Phi_0}{2\pi} \frac{\partial \phi}{\partial t}. \tag{2.24}$$

The voltage across a JJ depends upon the time derivative of the phase difference ϕ, while the current exhibits a sinusoidal dependence on ϕ. A JJ biased by a constant DC current, where $I < I_C$, exhibits a constant phase $\phi = \arcsin(I/I_C)$ where the voltage across this junction is zero. This situation corresponds to steady state DC operation in superconductive circuits.

If a constant voltage V_0 is applied across a Josephson junction, the phase of the JJ continuously changes. From (2.24),

$$\frac{\partial \phi}{\partial t} = \frac{2e V_0}{\hbar}, \tag{2.25}$$

$$\phi = \frac{2eV_0}{\hbar}t. \tag{2.26}$$

Substituting (2.26) into (2.23),

$$I(t) = I_c \sin\left(\frac{2eV_0}{\hbar}t\right). \tag{2.27}$$

A constant voltage applied across a JJ therefore produces current oscillating at a frequency

$$f = \frac{2e}{h}V_0, \tag{2.28}$$

where $\frac{2e}{h} \approx 483.6\,\text{GHz/mV}$. This phenomenon is known as the AC Josephson effect. A Josephson junction is therefore a natural voltage-to-frequency converter.

2.4.2 Josephson Inductance

A Josephson junction is a highly nonlinear circuit element. For small changes in current or voltage, however, a JJ can be treated as a nonlinear inductor. Consider a small change in current δ_I producing a phase change δ_ϕ.

$$I + \delta_I = I_c \sin(\phi + \delta_\phi), \tag{2.29}$$

$$\delta_I = I_c \cos(\phi)\delta_\phi, \tag{2.30}$$

$$U = \frac{\hbar}{2e}\frac{\partial \phi + \delta_\phi}{\partial t} = \frac{\hbar}{2e\,I_c \cos(\phi)}\frac{\partial \delta_I}{\partial t} = L_J\frac{\partial I}{\partial t}, \tag{2.31}$$

where L_J is the Josephson inductance of a junction,

$$L_J = \frac{\hbar}{2eI_c \cos(\phi)} = \frac{\Phi_0}{2\pi I_c \cos(\phi)} = \frac{L_{J0}}{\cos(\phi)}, \tag{2.32}$$

$$L_{J0} = \frac{\hbar}{2eI_c}. \tag{2.33}$$

This inductance depends upon the critical current of a JJ, which for a specific fabrication process is typically determined by the physical area. The inductance also depends on the phase difference, which can be controlled by a DC bias current.

2.4.3   Josephson Energy

The energy of a Josephson junction is

$$E_J = \int_0^{t_0} IV dt. \tag{2.34}$$

Based on the Josephson relations, (2.23) to (2.24),

$$E_J = \int_0^{t_0} I_c \sin(\phi) \frac{\Phi_0}{2\pi} \frac{\partial \phi}{\partial t} dt, \tag{2.35}$$

$$E_J = \frac{\Phi_0 I_c}{2\pi} \int_0^\phi \sin(\phi) d\phi, \tag{2.36}$$

$$E_J = \frac{\Phi_0 I_c}{2\pi} (1 - \cos(\phi)). \tag{2.37}$$

E_J is the potential energy accumulated by a JJ, similar to an inductor.

2.4.4   JJ Circuit Models

In a Josephson junction operating at a nonzero temperature or voltage, quasiparticles tunnel along with the Cooper pairs through the insulating barrier. These quasiparticles are due to the thermal breakup of Cooper pairs. The Josephson relations, (2.23) and (2.24), only describe the superconductive component of the total current. For the flow of quasiparticles, the JJ corresponds to a nonlinear conductance G_N which depends upon the voltage and temperature,

$$G_N = G(V, T). \tag{2.38}$$

A common approximation of the nonlinear conductance G_N is a piecewise function,

$$G_N = \begin{cases} \frac{1}{R_{SG}} & |V| \le \frac{2\Delta}{e}, \\ \frac{1}{R_N} & |V| \ge \frac{2\Delta}{e}. \end{cases}$$

R_N is the normal resistance corresponding to the breakup of Cooper pairs with energies above the energy gap, R_{SG} is the subgap resistance corresponding to the thermally excited quasiparticles, and Δ is the energy gap, as discussed in Sect. 2.3.3.

This simple model describing a JJ is the resistively shunted junction (RSJ) model [75] and is applicable to a DC-biased junction. The current-phase relation for the

RSJ model is

$$I = I_c \sin(\phi) + \frac{\Phi_0}{2\pi} G_N \frac{d\phi}{dt}. \tag{2.39}$$

This expression is the sum of the supercurrent and the resistive quasiparticle current.

As a JJ consists of two electrodes separated by an insulating barrier, the JJ also exhibits a certain capacitance. As in a standard capacitor, this junction capacitance affects the behavior of the JJ when the voltage across the JJ changes, producing a displacement current. In general, the junction capacitance depends upon the physical area and the thickness of the tunneling barrier.

The circuit model of a JJ, where a Josephson element, resistor, and capacitor are connected in parallel, is a resistively and capacitively shunted junction (RCSJ) model [75]. The total current through a JJ is the sum of the individual current components,

$$I = I_c \sin(\phi) + \frac{\Phi_0}{2\pi} G_N \frac{d\phi}{dt} + C \frac{dV}{dt}. \tag{2.40}$$

The RCSJ circuit model, schematically shown in Fig. 2.7, is a commonly used circuit model of a Josephson junction [79]. This model provides reasonable accuracy while maintaining high computational efficiency.

2.4.5 Dynamics of Josephson Junctions

Based on the RCSJ circuit model, an expression characterizing the current-phase relation of a JJ is

$$I = I_c \sin(\phi) + \frac{\Phi_0}{2\pi} G_N \frac{d\phi}{dt} + C \frac{\Phi_0}{2\pi} \frac{d^2\phi}{dt^2}. \tag{2.41}$$

This expression is a nonlinear differential equation with nonlinear coefficients. To simplify the expression, the conductance G_N is often assumed to be constant.

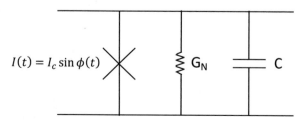

Fig. 2.7 Equivalent circuit of the RCSJ model of a JJ

$$I(t) = I_c \sin \phi(t)$$

Expression (2.41) is similar to the equation of motion of a particle with a certain mass M and damping η on a tilted washboard potential [79]. M is proportional to the junction capacitance C, and η is proportional to G_N. The tilt of the washboard is proportional to the current I. This dependence is depicted in Fig. 2.8.

An alternative analogy for describing the dynamic behavior of a JJ is a pendulum system [79]. Expression (2.41) corresponds to a damped mechanical pendulum of length l and mass M, deflected by an angle ϕ from normal. This pendulum is depicted in Fig. 2.9. The angle ϕ corresponds to the phase of a JJ, while the mass

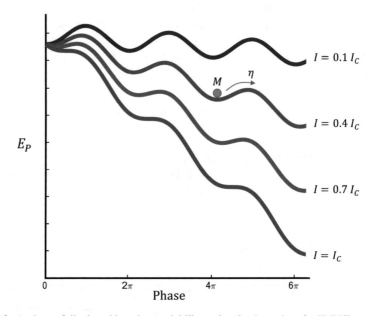

Fig. 2.8 Analogy of tilted washboard potential illustrating the dynamics of a JJ. Different curves correspond to different bias currents as compared to the critical current I_C

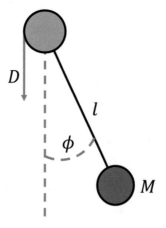

Fig. 2.9 Analogy of mechanical pendulum illustrating the dynamics of a JJ

and length correspond to the critical current of a junction. A torque D applied to a pendulum corresponds to the bias current of a JJ, rotating the pendulum and changing the phase ϕ. Damping of the pendulum is equivalent to the junction conductance, while the moment of inertia is equivalent to the junction capacitance. This analogy can be used to illustrate different kinds of dynamic behaviors of a JJ. A large torque rotates the pendulum (a continuous change in phase ϕ produces a nonzero voltage), while a fast kick of the pendulum rotates the pendulum once (a 2π change in phase produces a voltage pulse).

To simplify the process of characterizing a JJ, the Stewart-McCumber parameter β is used [80, 81], where

$$\beta = \frac{2e}{\hbar} I_C R_N{}^2 C. \tag{2.42}$$

Junctions with $\beta \approx 1$ are considered critically damped and are the primary type of JJ in superconductive digital electronics [75].

Junctions with $\beta \ll 1$ are overdamped and exhibit either a small capacitance or resistance. In the washboard potential analogy, an overdamped JJ corresponds to a particle with either a small mass or large damping. Consider a large tilt (current I), where a particle rolls down the washboard. If the tilt is slightly reduced, the particle quickly stops moving and is trapped in one of the nearby local minima.

Junctions with $\beta \gg 1$ are underdamped and exhibit either a large capacitance or resistance. In the washboard potential analogy, an underdamped JJ corresponds to a particle with either a large mass or small damping. This particle does not immediately stop moving once the tilt is reduced, and continues to roll down the washboard. To stop this particle from moving, the tilt (current I) has to be significantly decreased.

These properties produce two types of current-voltage characteristics for JJs. These two I-V characteristics are illustrated in Fig. 2.10. Overdamped JJs are non-hysteretic, and underdamped JJs are hysteretic. Current below I_C produces a zero voltage across the JJ, while current above I_C produces a finite voltage. In the overdamped case, if the current is decreased below I_C, the junction returns to the zero voltage state. In the underdamped case, however, a decrease in current below I_C produces a small but nonzero voltage.

2.5 Superconductive Devices

In this section, certain other important superconductive electronic devices are described, and applications of these devices to modern superconductive circuits are discussed.

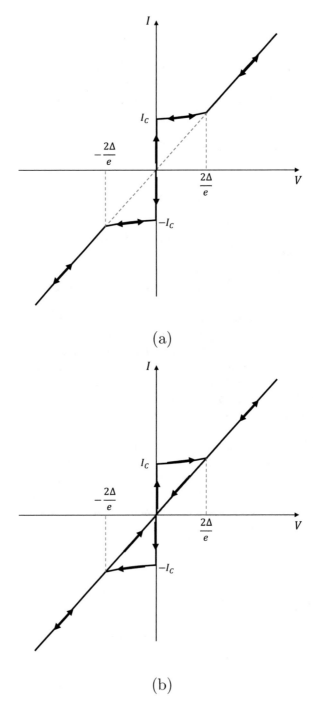

Fig. 2.10 Current-voltage characteristics of a JJ: (**a**) overdamped and (**b**) underdamped

2.5.1 Superconductive Nanowire Single-Photon Detectors

A superconductive nanowire single-photon detector (SNSPD) is a device capable of detecting a single incident photon [38]. This device consists of a long, narrow superconductive nanowire biased close to the critical current. An incident photon transfers energy to the electrons, breaking Cooper pairs and creating a local hotspot. This hotspot exhibits a comparably large resistance, producing a voltage pulse which can be detected and processed.

An example layout of an SNSPD is depicted in Fig. 2.11. A meandering structure increases the length of the nanowire and hence the area for detection, as any location on the nanowire is sensitive to photons. SNSPDs are widely used in quantum computing, photonics, detection, and communication circuits [83].

2.5.2 Cryotron

A cryotron is the first superconductive switch, proposed by D. Buck in 1956 [5]. It does not utilize the Josephson effect and is based on exploiting the critical field within a superconductor.

A cryotron, depicted in Fig. 2.12, is a four-terminal device, consisting of two superconductive wires. One wire is wrapped around another wire, and these wires are galvanically isolated. A control current I_C is passed through the coiled wire. This structure produces a magnetic field which disrupts the superconductive behavior in a nearby wire by exceeding the critical magnetic field. The device therefore operates as a switch, in which a comparatively small current I_C in a coil controls a larger current I_S in a straight wire.

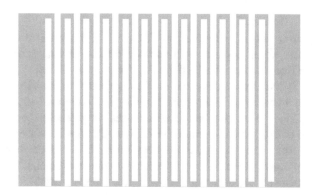

Fig. 2.11 Typical structure of a superconductive nanowire single-photon detector [82]

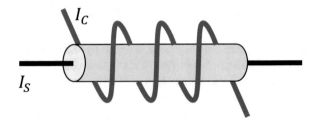

Fig. 2.12 Cryotron

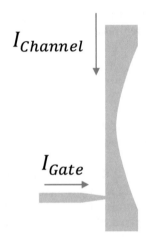

Fig. 2.13 Nanocryotron (nTron)

2.5.3 Nanocryotron

The nanocryotron (nTron), proposed in 2014, is a three-terminal electrothermal device [84]. The nTron consists of a superconductive channel and a weak link (choke). The nTron is schematically depicted in Fig. 2.13.

The nTron is initially supplied with a bias current through a superconductive channel. When a current pulse is applied to the gate, connected to a weak link (a constriction in the wire), a localized hotspot causes the channel to become resistive. This transition diverts the bias current into a load. The device is therefore capable of providing current gain, as a small gate current controls a large channel current. The nTron cools down, restoring superconductivity in the channel, diverting current back into the channel, and resetting the device to the initial state [47].

A nTron device is similar in operation to an SNSPD [38], as previously described. In an SNSPD, incident photons create localized hotspots, allowing current to be diverted into the load, producing a voltage.

2.5.4 Superconductor-Ferromagnetic Devices

A promising family of devices for superconductive electronics is based on the interaction between superconductive and ferromagnetic materials. These devices are typically composed of a different number and configuration of superconductive, normal metal, ferromagnetic, and insulating layers, and number of terminals [85, 86]. The primary objective of these devices is to provide signal gain and isolation between the input and output.

Although superconductivity and ferromagnetism are incompatible in bulk materials, in thin film devices these phenomena produce a proximity effect, where one layer affects the behavior of the other layers. In a superconductor/ferromagnetic thin film stack, superconductive electrons exhibit a very small penetration depth within a ferromagnetic layer. This influence of the ferromagnetic material layers enables asymmetric control of the device, as only regular quasiparticle current affects the behavior of the other superconductive layers within a stack.

A variety of different devices utilizing this effect have been proposed and fabricated [85, 86]. One of these devices, a three-terminal superconductor-ferromagnetic transistor (SFT) [86], is used here to illustrate the behavior of superconductive-ferromagnetic devices.

2.5.4.1 Superconductor-Ferromagnetic Transistor

The three-terminal SFT device is depicted in Fig. 2.14. This device consists of two junctions stacked above each other. The acceptor junction consists of an insulating layer (I) sandwiched between two superconductive layers (S), forming an SIS structure. The injector junction consists of an insulating layer between two ferromagnetic layers (F) and two superconductive layers, forming an $SFIFS$ structure. Both junctions share a superconductive layer, forming an $SFIFSIS$ device, where an SIS acceptor is stacked on top of an $SFIFS$ injector.

Operation of the SFT device is similar to previously proposed superconductive multilayered stacks, such as a quiteron [87], consisting of an $SISIS$ multilayer.

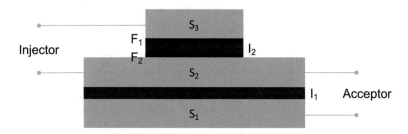

Fig. 2.14 Structure of a three-terminal SFT device. The superconductor layers are marked as S_1, S_2, and S_3, ferromagnetic layers as F_1 and F_2, and insulator layers as I_1 and I_2

In an $SISIS$ device, when one SIS junction is biased, excess quasiparticles are injected into the shared superconductive middle layer. This injection suppresses superconductivity within this layer, changing the properties of the second SIS junction. One important distinction, however, is the presence of ferromagnetic layers within the stack. These layers suppress Josephson current through one of the two junctions, making the device asymmetric, introducing input-output isolation [88].

In a three-terminal SFT, the current in the $SFIFS$ injector I_i introduces excess quasiparticles in the middle S_2 layer shared between the acceptor and injector, as well as the S_1 layer. This effect suppresses the superconductive energy gap Δ_1 and Δ_2 in, respectively, the S_1 and S_2 layers, reducing the critical current I_c of the acceptor SIS junction [45, 88]. Experimental behavior describing the $SFIFS$ injector, such as the linear I-V characteristics of the stack, suggests that the injector current I_i does not exhibit Josephson behavior due to the presence of an exchange field in the ferromagnetic layers [88, 89]. Current-controlled modulation of the critical current along with good input-output isolation is the primary advantage of SFT devices as compared to earlier similar structures [86].

2.6 Conclusions

The basic properties of superconductors are discussed in this chapter. The primary theoretical framework for the analysis of low temperature superconductive materials—the London, Ginzburg-Landau, and Bardeen-Cooper-Schrieffer theories—is described. The defining features of superconductive materials are discussed, along with different types of materials and characteristics. The primary properties of these materials are emphasized in relation to superconductive electronics. The properties and dynamic behavior of Josephson junctions are also discussed with both intuitive analogies describing the dynamic behavior and classic circuit models. Important cryogenic devices commonly used in superconductive electronics are also briefly reviewed.

Chapter 3
Superconductive Circuits

Superconductive digital and analog circuits are introduced in this chapter. The use of superconductive devices to build electronic circuits was first proposed with the invention of the cryotron in the 1950s [5], as described in Chap. 2. In a cryotron, a magnetic field produced by a current in one wire switches another wire between the superconductive and normal states. As a four-terminal switch with isolated control and channel currents, a cryotron can be used to construct flip flops and logic gates [90]. With the advent of transistor-based integrated circuits capable of operating at room temperature, cryotron-based electronics were no longer a competitive technology. Practical research and development efforts in superconductive electronic circuits transitioned to specialized applications.

One of these applications is sensing small magnetic fields. Superconductive quantum interference devices (SQUIDs) are Josephson junction-based devices capable of sensing extremely small (up to 5×10^{-18} T) magnetic fields [37]. Certain variations of SQUIDs are also the essential component of many superconductive digital circuits. In Sect. 3.1, SQUID and related applications are described.

Research on applying superconductive devices to digital electronic applications continued in IBM [18]. Fast (on the order of picosecond) switching speeds of Josephson junctions were exploited to achieve multi-gigahertz system clock frequencies. Apart from switching speeds, other benefits of JJ-based circuits as compared to semiconductor circuits are low power dissipation, extremely small thermal noise, and zero DC interconnect resistance. IBM efforts focused on voltage level logic [18], where information is represented, similar to conventional semiconductor electronics, in the form of different voltage levels. In Sect. 3.2, latching voltage logic, developed at IBM, is described.

Another approach to Josephson logic, rapid single flux quantum (RSFQ) logic, was developed in 1985 in Moscow State University [91]. In this logic family, information is represented by small voltage signals with a quantized area—single flux quantum (SFQ) pulses—rather than the voltage level. This approach produces multiple benefits for high speed operation of superconductive circuits. In Sect. 3.3,

© Springer Nature Switzerland AG 2022
G. Krylov, E. G. Friedman, *Single Flux Quantum Integrated Circuit Design*,
https://doi.org/10.1007/978-3-030-76885-0_3

RSFQ logic is briefly introduced. RSFQ logic is described in much greater detail in Chap. 4.

Reciprocal quantum logic (RQL) is an alternative SFQ logic family [92] utilizing AC bias currents rather than the DC bias currents used in RSFQ. Information in RQL is represented as the presence or absence of a pair of SFQ pulses. Multiple advantages and disadvantages of RQL logic exist as compared to RSFQ. This circuit family and related issues are described in Sect. 3.4.

A major advantage of superconductive circuits for energy efficient computing is the capability of adiabatic operation. In adiabatic circuits, transitions between states are gradual, and the energy dissipation of these circuits approaches zero for sufficiently slow transitions. In Sect. 3.5, a brief introduction to one type of adiabatic superconductive logic—quantum flux parametron (QFP) logic [24]—is provided.

Any computing technology requires memory for storing instructions and computational results. Efficient and dense memory remains to this day an open problem in superconductive circuits. In Sect. 3.6, memory suitable for use in conjunction with cryogenic logic is described. Finally, in Sect. 3.7, a brief summary is provided.

3.1 SQUID

In this section, one of the first practical superconductive circuits, the superconductive quantum interference device (SQUID), is introduced and described. To date, SQUID-based detectors are the most sensitive detectors of magnetic fields [93]. Two different types of SQUID topologies are discussed here—the single junction SQUID in Sect. 3.1.1 and the two junction SQUID in Sect. 3.1.2.

3.1.1 Single Junction SQUID

A single junction SQUID, also commonly referred to as an RF SQUID, was invented in 1965 by A. Silver and J. E. Zimmerman [94]. This device exploits the interference of the superconducting wave function across a Josephson junction.

Consider the circuit shown in Fig. 3.1—a superconductive loop interrupted by a Josephson junction. In a superconductive loop, the magnetic flux is quantized,

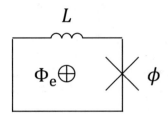

Fig. 3.1 RF SQUID

as discussed in Sect. 2.3.2. If the loop is interrupted by a JJ, however, the flux is typically not quantized [95]. The voltage across a JJ can only exist due to a change in the magnetic flux within a loop [75],

$$V = \frac{d\Phi}{dt}. \tag{3.1}$$

From the Josephson relations, (2.23) and (2.24),

$$\frac{d\phi}{dt} = \frac{2\pi}{\Phi_0} V. \tag{3.2}$$

Substituting (3.1) into (3.2) and integrating over time,

$$\phi = 2\pi \frac{\Phi}{\Phi_0}, \tag{3.3}$$

assuming the integration constant is zero. The phase difference ϕ across the junction is therefore periodic with respect to the internal magnetic flux Φ.

Substituting (3.3) into the Josephson current-phase relation,

$$I = I_c \sin \phi = I_c \sin \left(2\pi \frac{\Phi}{\Phi_0} \right), \tag{3.4}$$

where I_c is the critical current of the JJ. The internal flux consists of two parts—an externally applied flux Φ_e and the flux produced by the loop current induced by this external field,

$$\Phi = \Phi_e - LI. \tag{3.5}$$

Substituting (3.5) into (3.4),

$$I = I_c \sin \left(2\pi \frac{\Phi_e}{\Phi_0} + 2\pi \frac{LI}{\Phi_0} \right), \tag{3.6}$$

or

$$I = I_c \sin \left(2\pi \frac{\Phi_e}{\Phi_0} + 2\pi \lambda \frac{I}{I_c} \right), \tag{3.7}$$

where

$$\lambda = 2\pi \frac{LI_c}{\Phi_0} \tag{3.8}$$

is the primary parameter characterizing a SQUID, which determines the shape of the dependence of Φ on Φ_e.

Expression (3.7) shows that the current in a SQUID is periodic with respect to both an externally applied flux and an applied bias current. The periodic nature of the interference pattern enables multiple useful applications of a single junction SQUID. Among these applications are logic circuits [96], A/D converters [35], and sensitive circuits for measuring magnetic flux, the RF SQUID [97].

3.1.2 Two Junction SQUID

The two junction SQUID was invented in 1964 by R. C. Jaklevic, John Lambe, A. H. Silver, and J. E. Mercereau [98]. The circuit consists of a superconductive loop interrupted by two Josephson junctions connected in parallel, as shown in Fig. 3.2.

For this structure, an expression for the total flux through a loop similar to (3.5) is [75]

$$\Phi = \Phi_e + L_1 I_1 - L_2 I_2, \tag{3.9}$$

$$I = I_1 + I_2, \tag{3.10}$$

where I is the total current within the loop. Similar to (3.3), the difference between the JJ phases in this structure is

$$\delta\phi = \phi_1 - \phi_2 = 2\pi \frac{\Phi}{\Phi_0}. \tag{3.11}$$

The external magnetic flux in a two junction SQUID changes the current-voltage characteristics of the device, enabling precise measurement of the magnetic flux from the DC bias current.

A two junction DC SQUID is biased by a DC current, which is initially equally divided between the two branches. When a small magnetic flux ($\Phi < \frac{\Phi_0}{2}$) is applied to a device, a screening current I_s circulates within the loop to bring the total flux through the loop to the nearest integer—in this case, to cancel the applied flux. This response changes the current distribution between the branches—the current in one branch is increased by I_s, while the current in another branch decreases by I_s. If either of these currents exceeds the critical current I_c of the JJ in the corresponding branch, a voltage V is produced across the device.

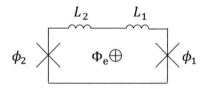

Fig. 3.2 Two junction SQUID

If the applied flux exceeds $\frac{\Phi_0}{2}$, the circulating screening current changes direction to bring the total flux to the nearest integer—in this case, to Φ_0. This behavior produces a periodic dependence of the branch currents on the applied flux with a period of Φ_0.

If the bias current supplied to the device exceeds $2I_c$, the device operates in the resistive mode. In this case, a voltage is always present across the device and changes with the applied flux with a period of Φ_0.

As with a single junction SQUID, the periodic dependence of voltage on the applied magnetic flux and current can be used for sensitive detectors of magnetic flux [97] and within A/D converters [35]. Moreover, the dependence of the device current-voltage (I-V) characteristics on the applied flux is used in both latching voltage state superconductive electronics [99] and single flux quantum memory [100].

3.2 Voltage Level Logic

Voltage level Josephson logic, invented and developed in IBM since 1964 [18], is briefly described in this section. In 1967, J. Matisoo proposed a tunneling cryotron—a device similar to a standard cryotron—that utilizes a Josephson junction as a gate. This development enabled the first Josephson junction-based processors. Efforts at IBM concentrated on voltage level circuits, where information is represented as discrete voltage levels, similar to conventional CMOS circuits [101].

Multiple types of voltage level superconductive circuits exist [102, 103]. In these circuits, the gates are typically based on multi-junction SQUID loops. An example circuit of a typical voltage level OR gate is schematically shown in Fig. 3.3. To produce a DC voltage, a hysteretic Josephson junction is switched into the voltage state. This hysteretic switching exploits the inherent latching property of these circuits. To reset the state of the logic gates into the initial state, the bias current of every combinatorial network is decreased below a specific return current during each clock cycle. The critical currents and inductances within the SQUID loops are tuned to switch from the superconductive state into the voltage state of the SQUID when a specific control current is applied. This signal voltage is transferred between gates by matched superconductive transmission lines.

Josephson logic enabled sub-nanosecond switching times as early as 1967 [90], incentivizing further development of Josephson technology for high speed computing. Multiple obstacles, however, existed in developing a large scale logic circuit. One of the primary issues is the latching nature of this logic. A high frequency circuit in this technology requires a high frequency AC bias network, which is a complex system to design and manufacture [23]. Furthermore, latching circuits sometimes fail to reset if the rate of change in this AC bias/clock signal is high [104]. This phenomenon is called punchthrough, where a nonzero probability

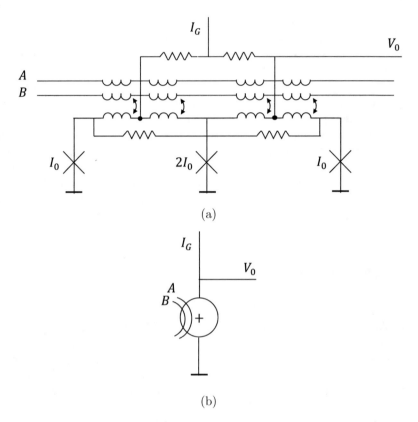

(a)

(b)

Fig. 3.3 Voltage level OR gate [102], (**a**) circuit schematic, and (**b**) logic symbol

exists even if the rate of change is low [105]. Another problem with these early JJ-based circuits was the lead alloys used during the fabrication process [23]. Lead alloys experience progressive degradation during thermal cycling, which is unavoidable in superconductive circuits, affecting the circuit characteristics.

However, the primary reason for terminating the IBM effort in 1983 was due to major advancements in silicon transistor technology [18]. Superconductive circuits still exhibited performance advantages as compared to silicon electronics, but the maximum clock frequencies achievable by voltage level superconductive logic were limited to a few Ghz [23]. These frequencies were achievable in room temperature silicon and gallium arsenide technologies without requiring cryogenic operation [106]. The performance benefits of superconductive logic circuits were insufficient to tolerate the significant disadvantage of cryogenic operation.

3.3 RSFQ Logic

The rapid single flux quantum (RSFQ) [23] logic family for superconductive electronic circuits is introduced in this section. RSFQ is a logic family for low power, high performance cryogenic computing based on Josephson junctions, first introduced in 1985 by K. Likharev, V. Semenov, and O. Mukhanov [91]. RSFQ circuits are among the fastest digital circuits. An RSFQ T flip flop has been demonstrated to operate at 770 Ghz [34].

As opposed to voltage level logic, the JJs used in RSFQ circuits are damped close to the critical damping level ($\beta_c \sim 1$) while maintaining a non-hysteretic current-voltage characteristic. Non-hysteretic JJs naturally produce a voltage pulse, called a single flux quantum (SFQ) pulse. Each SFQ pulse corresponds to a shift in the superconductive phase difference across a JJ by 2π—an event referred to as switching a JJ. These voltage pulses exhibit a quantized area of $\Phi_0 \approx 2.07$ mV·ps. A properly biased and damped JJ in RSFQ technology reproduces a single SFQ pulse before returning to the superconductive state, eliminating the need to reduce the bias current. This behavior is schematically shown in Fig. 3.4. In RSFQ technology, binary information is represented as the existence or absence of an SFQ pulse during a specific clock period, respectively, a logic one state or logic zero state.

During early development of RSFQ technology, logic gates were connected by resistors. RSFQ was therefore abbreviated as "resistive single flux quantum." These resistors have since been replaced by small inductors, and the RSFQ acronym became "rapid single flux quantum." All RSFQ circuits require a bias current; however, as opposed to an RF current in voltage level logic, RSFQ circuits are biased by a DC current [49]. In conventional RSFQ circuits, resistors are utilized within the bias distribution network [23]. Alternative bias schemes have been proposed. A more complete discussion of RSFQ logic gates, transmission lines [107], and bias schemes [50] is provided in Chap. 4.

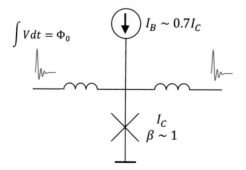

Fig. 3.4 Basic RSFQ circuit—transmission line

3.4   Reciprocal Quantum Logic

Reciprocal quantum logic (RQL) was proposed by Q. Herr in Northrop Grumman in 2011 [92]. In this technology, reciprocal pairs of single flux quantum pulses (positive and negative) represent a logic "one," while the absence of a pair of pulses represents a logic "zero." RQL circuits utilize an AC bias/clock supply, and individual gates are coupled to these AC power lines by transformers.

RQL gates are composed of transformers, JJs, and inductors. An example of an RQL circuit—a transmission line—is shown in Fig. 3.5. An AC clock is coupled to the gates by transformers. The basic logic elements in this technology are A-AND-NOT-B and AND-OR gates, combined with RS flip flops [92].

An AC clock with a single phase does not provide directionality to the signals—any forward propagation of a signal during the positive half cycle is compensated by backward propagation during the negative half cycle [108]. Multiple clock phases are therefore necessary to propagate data throughout a circuit. As RQL gates are biased by the same AC clock, these circuits require a multiphase AC power distribution network. RQL provides a natural solution to uncertainty in gate timing—the data pulses are synchronized by each individual clock phase, exhibiting a self-correcting timing scheme [92].

The primary advantage of RQL is low power dissipation within the cryogenic environment—the AC power lines are terminated off-chip. Another important advantage is the serial application of the bias currents. Conventional RSFQ circuits are biased in parallel. For large scale RSFQ circuits, these currents can exceed many amperes, greatly complicating the design of the bias lines and ground planes [110]. In RQL, the gates are coupled to the same power bus, greatly reducing the magnitude of the required current. An example of a multiphase RQL clock/bias scheme is shown in Fig. 3.6.

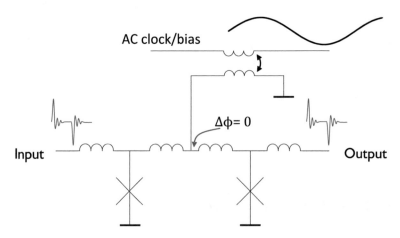

Fig. 3.5 Reciprocal quantum logic transmission line [108]

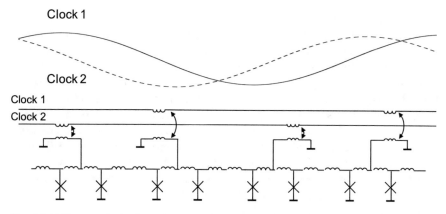

Fig. 3.6 RQL transmission line with a typical clock/bias topology [109]

Among the disadvantages of RQL circuits is the aforementioned multiphase power bias system. Multi-gigahertz AC power distribution is a complex task. Superconductive power dividers occupy a significant fraction of the circuit area in RQL circuits [111], comparable to a RSFQ clock network. RQL gates are also primarily composed of transformers, which are difficult to scale in current superconductive circuit fabrication technologies [112].

3.5 Adiabatic Superconductive Logic

In thermodynamics, adiabatic processes do not change the total heat of a system. Adiabatic logic circuits are low power circuits that operate in a reversible manner, dissipating little heat. In this section, the general principles of reversible computing are introduced, and a superconductive circuit topology capable of reversible operation—quantum flux parametron (QFP)—is described.

3.5.1 Reversible Computing

A theoretical limit on energy dissipation in computing systems exists. Landauer's principle, named after R. Landauer who first discovered this limit [113], states that only the erasure of information dissipates energy. This concept is a consequence of the laws of thermodynamics. Computation decreases the total entropy of a system. This decrease in system entropy has to be compensated by an increase in physical entropy, which is eventually dissipated as heat. The minimum energy required to erase one bit of information is $kT \ln 2$, where k is Boltzmann's constant and T is the temperature. This energy, sometimes called the Landauer limit [114], is 2.9×10^{-21} J at room temperature (300 K), and 4×10^{-23} J at 4.2 K, a standard temperature for superconductive electronics. Although this energy is much lower than the energy dissipated in modern circuits, the Landauer limit poses a fundamental limitation on energy efficiency.

Reversible computation is the computational process that can be reversed in time. Two types of reversibility exist—logical and physical. Logically reversible computation consists of reversible operations. Given an output of a reversible logic computation, the input can in principle be recovered, thereby undoing the computation. An example of a reversible logical operation is inversion [115]. Given an output of inversion, the input can be obtained. An AND gate, alternatively, is irreversible—given an output, it is not possible to recover the inputs. A set of reversible logical operations exists. These operations all exhibit an equal number of inputs and outputs to preserve information.

Physically reversible computations produce logically reversible operations [113] in a physically reversible way, preserving the capability to undo the operation. The Landauer limit does not apply for a physically reversible computation as information is not erased. This property enables computing with extremely low energy dissipation.

Many different circuit realizations of adiabatic computing have been proposed. Superconductive circuits are among the most promising of these circuits, as the dynamic energy dissipation in these circuits is extremely small. A common superconductive logic family for adiabatic and reversible computing is the adiabatic quantum flux parametron (AQFP) logic family [24].

3.5.2   Adiabatic Quantum Flux Parametron

The quantum flux parametron was proposed by E. Goto and researchers from the University of Tokyo in 1987 [116]. This technology does not utilize single flux quantum pulses. Rather, trapezoidal current waveforms are used.

The fundamental quantum flux parametron device is shown in Fig. 3.7. The device consists of two superconductive loops, $J1 - L1 - L_q$ and $J2 - L2 - L_q$, capable of storing a single magnetic flux quantum. QFP circuits are clocked by a multiphase AC clock signal, similar to RQL circuits. I_{in} is a small input current, corresponding to a data bit. An AC excitation clock signal I_x induces magnetic flux within the device. Depending upon the direction of I_{in}, a single flux quantum is stored in either the left or right loop of the device. A comparatively large output current is generated across L_q, where the direction of the current corresponds to the logic state. An important feature of a QFP device is that the output current only depends on the direction of the input current, not the magnitude [117]. This device therefore exhibits high current gain.

AQFP logic primarily utilizes majority gates. Combined with inverters, which in this technology do not require additional area (only a change in the transformer

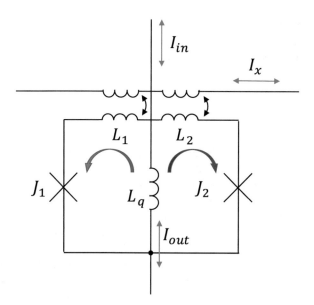

Fig. 3.7 Quantum flux parametron device [24]

polarity), these gates form a complete logic set. Depending upon the junction damping parameters, the transition between the states of the device can occur without a sudden change in the phase of the JJs. This gradual change in the device state enables reversible adiabatic operation [117]. AQFP circuits demonstrate five orders of magnitude lower energy-delay product (EDP) as compared to 12 nm CMOS circuits [118]. Including the energy for cryogenic cooling, AQFP circuits exhibit a two orders of magnitude EDP advantage [118].

This technology is also capable of both physical and logical reversible operations, with an energy per gate of approximately 2×10^{-23} J [36]—below the Landauer limit, as information is not erased in the computational process.

Among the disadvantages of AQFP is the difficulty in distributing a multiphase AC clock. As this technology is primarily aimed at highly energy efficient computation, the clock frequency of AQFP circuits is generally lower than encountered in RQL and RSFQ circuits, partially alleviating the issue of complex multiphase AC clock delivery.

3.6 Memory

All computing technologies require memory to store states and the results of computation. In this section, different types of memory suitable for use within a cryogenic environment and capable of interfacing with superconductive electronics are reviewed. In Sect. 3.6.1, a JJ-based approach for superconductive memory is

described. In Sect. 3.6.2, a cryogenic CMOS memory is discussed along with the
necessary interface circuits. In Sect. 3.6.3, a prospective spin-based memory for
superconductive circuits is described.

3.6.1 Josephson Memory

Storage of memory states in the form of a circulating current [119] or the presence
of a magnetic flux quantum within a superconductive loop [100] was proposed
in the 1960s and 1970s before the development of single flux quantum logic.
Despite recent advances in the field of superconductive logic, dense and efficient
superconductive memory remains an open problem. In this subsection, the most
compact memory topology among the proposed structures [112], the nondestructive
vortex transitional (VT) memory, is described [120].

   The VT memory cell consists of two superconductive loops [112], storage loop
$J1 - L1 - L2 - J2$ and sense SQUID loop $J3 - J4$, magnetically coupled to the
storage loop, as shown in Fig. 3.8. The cell can be accessed by coincident current
pulses on the IX and IY control inputs. Positive pulses produce a logic "zero" state
within a storage loop, while negative pulses produce a logic "one" state in the form
of magnetic flux within the storage loop.

   The cell is nondestructively read out by applying a positive read current pulse on
control line IS. In the case of a logic "one" state in the storage loop, this current

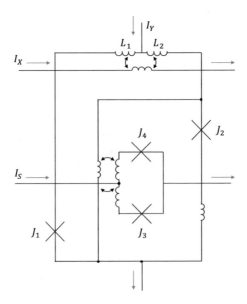

Fig. 3.8 Vortex transitional memory cell [112]

pulse causes the sensing SQUID to momentarily switch to a voltage state, producing an output pulse.

A 12 × 6 memory array structure composed of these cells has been manufactured with a density of 0.88 Mbit/cm^2 [112] using an advanced 600 µA/µm Niobium fabrication process [121]. Although this memory is currently the most dense Josephson memory array, this array remains significantly smaller than the density of modern CMOS memories [122]. Most of the area of these cells is occupied by transformers and inductors, which are difficult to scale [112].

3.6.2 Cryogenic CMOS Memory

Another option for dense memory in superconductive electronics is the cryogenic CMOS memory. In this subsection, a Josephson-CMOS hybrid memory is described.

A CMOS circuit exhibits multiple advantages when operated at cryogenic temperatures [123]. The carrier mobility of the transistors increases, the subthreshold slope becomes more steep, and the threshold voltage can be reduced, producing higher switching speeds and lower leakage currents [124]. One of the primary features of cryogenic CMOS memory, however, is the near infinite retention time of CMOS DRAM cells operating within a cryogenic environment [123]. This characteristic alleviates the need for a memory refresh sequence when operating within a cryogenic environment and enables the use of compact DRAM cells rather than SRAM cells.

Multiple issues with combining CMOS memory with RSFQ circuits exist. As RSFQ circuits are niobium based, this structure requires the CMOS memory to be placed on a separate IC, complicating the overall system design process. In addition, the interface between CMOS and SFQ is difficult due to the significant difference in magnitude of the signal voltages. CMOS circuits typically operate with an approximately one volt square waveform, while RSFQ circuits produce millivolt level pulses. This difference requires a complex combination of CMOS and SFQ amplifiers.

A common SFQ amplifier for interfacing with CMOS circuits is the Suzuki stack [125]. This amplifier is schematically shown in Fig. 3.9. The amplifier is composed of a stack of serially connected JJs separated into two parallel branches and biased close to the critical current. The bias current is initially equally split between the two branches of the stack. When an input pulse is applied, one of the JJs on the left side of the stack switches. This behavior increases the bias current on the right side, simultaneously switching all of the junctions on the right side. The bias current is redirected into the left branch of the stack, switching the remaining JJs.

Near simultaneous switching of many JJs produces a relatively high output voltage, on the order of 100 mV. This voltage can drive a CMOS amplifier, which in turn can drive a CMOS memory. A Suzuki stack is a latching element; the bias current returns to zero between the input pulses [126]. The stack is therefore

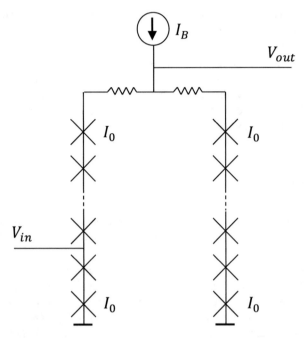

Fig. 3.9 Suzuki stack [126]

synchronized by a standard CMOS clock signal at the target frequency of the memory array.

To date, a 64 kb hybrid Josephson-CMOS memory has been demonstrated with an access time of 400 ps [127]. Although relatively slow as compared to RSFQ logic operation, this type of memory is a promising solution for RAM in RSFQ systems due to the high density and the capability for exploiting advances in CMOS fabrication technology.

3.6.3 Spin-Based Memory

Magnetic memory is currently widely considered as a nonvolatile replacement for several layers of the conventional CMOS memory hierarchy, such as SRAM and DRAM [128]. In this subsection, applications of this type of memory for superconductive electronics are reviewed.

Two prospective types of magnetic memory that operate within a cryogenic environment are cryogenic orthogonal spin transfer (COST) [129] and cryogenic spin hall effect (CSHE) [130] devices. These magnetic devices consist of a stack of different ferromagnetic and insulator materials, where the resistance of these devices is varied by changing the orientation of the magnetization of the different layers. COST devices utilize a spin-valve structure with a polarizer for fast and low energy switching of the magnetization of the free layer. The major advantage of COST devices, as compared to CSHE, is the low impedance of the magnetic memory element, providing enhanced compatibility with SFQ logic.

These spin-based devices can serve as memory elements; however, the devices require relatively high drive currents as compared to the currents typically used in SFQ circuits. As SFQ logic cannot directly supply these high currents, special driver devices, controlled by SFQ pulses, are necessary. A recently developed driver device, compatible with SFQ circuits and suitable for supplying high currents, is the nTron device [84]—an electrothermal switch that can be controlled by an SFQ pulse, as described in Sect. 2.5.3 [47].

A simple 4×4 memory array integrating CSHE MTJ devices with hTron elements (a variation of the nTron device) has recently been demonstrated [131]. Although currently in early development, this approach exhibits an important advantage—memory cells fabricated on the same IC as the logic circuits, alleviating the need for multiple ICs and frequent passing of data signals across long interconnect and temperature gradients.

3.7 Conclusions

Superconductive circuits are discussed in this chapter. Both analog and digital circuits are described, and memory topologies are presented. Among the analog devices, single and two junction SQUIDs are introduced along with characteristic expressions and common applications. Different families of superconductive digital logic, such as voltage level logic, rapid single flux quantum logic, reciprocal quantum logic, and adiabatic quantum flux parametron logic, are described. The basic principles of adiabatic and reversible computing are also reviewed. Finally, different types of cryogenic memory are introduced and the advantages and disadvantages of these memory topologies are discussed.

Chapter 4
Rapid Single Flux Quantum (RSFQ) Circuits

As briefly discussed in Chap. 3, rapid single flux quantum (RSFQ) [23] is a logic family targeting low power, high performance cryogenic computing. This logic family is based on Josephson junctions and was first introduced in 1985 [91]. Development of SFQ circuit fabrication and technology has enabled complex integrated circuits approaching 11,000 JJs for RSFQ digital signal processors [132] and similar complexity RSFQ prototype microprocessors [133–135]. These circuits operate at clock frequencies of tens of gigahertz. SFQ circuits with a regular layout structure, such as an AC-biased SFQ shift register, have been successfully manufactured and are used as fabrication process benchmarks. This shift register circuit is composed of over 800,000 JJs [136].

In this chapter, the RSFQ logic family for superconductive electronic circuits is described. Different transmission lines suitable to transfer picosecond SFQ signals are reviewed in Sect. 4.1. Key principles of RSFQ circuit topologies are introduced, and RSFQ logic gates and flip flops are discussed in Sect. 4.2. In Sect. 4.3, the distribution of bias currents in RSFQ circuits is reviewed, and energy efficient modifications of RSFQ logic—low voltage RSFQ, dual-rail SFQ, and ERSFQ—are discussed. A brief summary is provided in Sect. 4.4.

4.1 Transmission Lines

SFQ pulses are transferred across a circuit using two distinct types of transmission lines—active Josephson transmission lines (JTL) and passive transmission lines (PTL) [107]. Multiple advantages and disadvantages exist for each type of transmission line [137]. These structures and related features are reviewed in this section.

© Springer Nature Switzerland AG 2022
G. Krylov, E. G. Friedman, *Single Flux Quantum Integrated Circuit Design*,
https://doi.org/10.1007/978-3-030-76885-0_4

4.1.1 Josephson Transmission Lines

A JTL is a chain of grounded, shunted, and biased JJs connected in parallel by small inductors, as shown in Fig. 4.1. The bias currents and inductances are typically chosen to be equal to simplify the design and analysis process, although this requirement is not strict. The parameters of the individual JJs within a JTL can be tuned to achieve specific design objectives such as the propagation delay or peak voltage of an SFQ pulse.

The phase of the first JJ within a JTL changes by 2π upon the arrival of an input SFQ pulse. An SFQ voltage pulse produces an additional current Φ_0/L which is passed to the next Josephson junction within a JTL [138]. As this next junction is also biased, the total current momentarily exceeds the critical current of the JJ, placing the JJ into the normal (nonzero voltage) state, changing the JJ phase by 2π. This process regenerates and passes the SFQ pulse along a transmission line.

The inductance in a JTL is chosen to prevent storage of a flux quantum within a JTL loop. A rule of thumb for the JTL inductance for a target critical current is

$$L = \frac{\Phi_0}{2I_c}, \tag{4.1}$$

providing wide parameter margins (tolerance to parameter variations) and relatively low delay. The bias current is typically chosen to be approximately 70% of the JJ critical current [23].

The JJs within a JTL can be individually biased, but are commonly biased in pairs, as shown in Fig. 4.1. This approach reduces the number of bias elements and connections to the bias network.

JTLs are commonly used to connect individual SFQ logic circuits. The JTLs exhibit wide parameter margins and regenerate incoming SFQ pulses, providing noise discrimination. Among the disadvantages of using JTLs for interconnect are large area, limited routing distance between two JJs, and significant power dissipation in long lines composed of many JJs.

The routing distance between any two JJs in a JTL is constrained by the inductance,

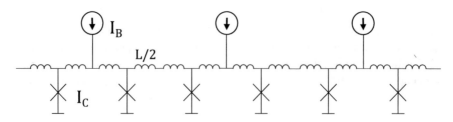

Fig. 4.1 Josephson transmission line

$$L < \frac{\Phi_0}{I_c}, \tag{4.2}$$

to prevent flux storage. For a JTL with a 250 μA junction, this restriction places a limit on the maximum inductance of ∼ 8 pH. The actual inductance is lower to consider parameter variations in the fabrication process. In modern superconductive fabrication processes (e.g., the 100 μA/μm² MIT Lincoln Lab SFQ5ee process [139]), this constraint limits the routing length between any two JJs to approximately 5 to 15 μm.

Each time a JJ switches within an RSFQ circuit, dynamic energy E_D is dissipated within the shunt resistor [140],

$$E_D \sim \int I V \, dt \sim I_C \int V \, dt = I_C \Phi_0, \tag{4.3}$$

where I_C is the critical current of the JJ. For a 250 μA junction, an energy $E_D \sim 5 \times 10^{-19}$ J is dissipated, or ∼ 0.5 nW per GHz. Although small, for a large high frequency system, this energy is a significant fraction of the overall power dissipated by the interconnect.

A JTL exhibits a relatively high delay—several picoseconds per JJ. Moreover, each JJ produces timing jitter—uncertainty in the switching time. With a large number of JTL stages, this uncertainty can accumulate and complicate the timing analysis process. JTLs, while efficient for routing over short distances, are therefore not practical for long-distance interconnect [107].

4.1.2 Passive Transmission Lines

SFQ pulses can propagate ballistically on superconductive striplines [141, 142]. This property enables the transfer of SFQ signals over long distances at the speed of light in the medium while exhibiting small degradation of the signal [143].

A typical PTL structure is shown in Fig. 4.2. A passive transmission line consists of a stripline or microstripline connected and impedance matched to a driver and receiver [144]. A small (∼ 0.5 Ω) resistor is typically introduced to disrupt the long superconductive loop formed between the driver and receiver to prevent current redistribution, reduce signal reflections, and decouple the driver from the receiver.

The primary function of the driver and receiver is to match the impedance of the cells to the impedance of the passive stripline, reduce signal reflections, and amplify degraded signals. Impedance matching in superconductive striplines is similar to the more general case of impedance matching in conventional transmission lines.

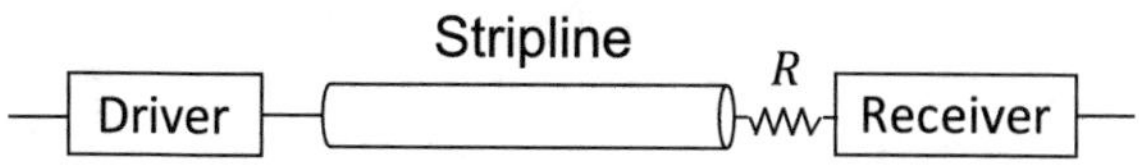

Fig. 4.2 Passive transmission line for SFQ signals

The input impedance of the receiver and output impedance of the driver, as well as the stripline characteristics, are adjusted to maximize the transfer of power and minimize the reflection coefficient [145]. Multiple topologies exist for the driver and receiver circuits [146]. Each topology requires different matching elements and number of Josephson junctions [147, 148].

The impedance of a PTL for a target fabrication process is inversely proportional to the width of the PTL. When routing large scale circuits, it is desirable to use narrow lines to increase the total available wiring resources. Narrow lines, however, exhibit a high impedance and are difficult to match to low impedance JJ-based circuits. PTLs therefore exhibit a tradeoff between wiring density (a function of line width) and the robustness of the driver/receiver.

PTLs exhibit multiple advantages for interconnect routing in large scale RSFQ circuits [107, 137]. Power is only dissipated in PTLs within the driver and receiver circuits. Due to the high speed of the signals within a PTL, most of the delay is produced by the driver and receiver, particularly in short lines. This situation makes a PTL less appropriate for short interconnect; there is simply too much overhead for the driver and receiver. In addition, PTLs are sensitive to noise, coupling, and impedance discontinuities in the stripline.

4.2 Logic Gates and Flip Flops

RSFQ logic gates are composed of different combinations of superconductive loops storing or not storing a quantum of magnetic flux depending upon the target function. These gates are typically clocked, where a logic one (zero) is represented as the presence (absence) of an SFQ pulse within a clock period [23]. The operation of commonly used RSFQ logic gates is described in this subsection.

4.2.1 D Flip Flop

A D flip flop (a DFF or, alternatively, an RS flip flop) with a destructive readout is one of the least complex RSFQ circuits. This circuit can be used to demonstrate certain important principles common to most RSFQ gates.

A D flip flop is schematically depicted in Fig. 4.3. It consists of a superconductive loop, $J3 - L - J2$ (similar to the SQUID loop described in Sect. 3.1.2), as shown in Fig. 4.4, and a clock input at junction J1. The inductance L is chosen to be sufficiently large to store a magnetic flux quantum, $L > \Phi_0/I_c$. A typical value of $L = 1.25\Phi_0/I_c$. For example, in a typical DFF, $L = 10$ pH and $I_C = 250\ \mu$A.

A DFF circuit operates as follows. Most of the bias current is initially directed into J3, as this path exhibits a smaller inductance. This state corresponds to logic zero. An incoming input SFQ pulse switches J3, redistributing the current while redirecting most of the bias current into J2. This state corresponds to logic one. Alternatively, this state can be represented as an equal distribution of bias

Fig. 4.3 RSFQ D flip flop

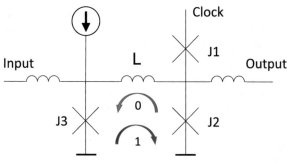

Fig. 4.4 Basic RSFQ storage loop

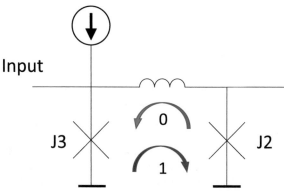

Fig. 4.5 RSFQ balanced comparator (decision making pair)

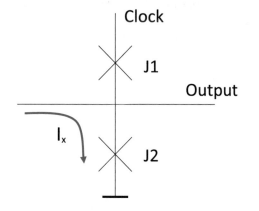

current between J2 and J3 with a circulating persistent current $I_s = \Phi_0/L$ changing direction depending upon the state.

The persistent current within a DFF storage loop is sustained indefinitely. To read the state of the flip flop, a clock/reset pulse is supplied to the clock input. This pulse switches either J1 or J2 depending upon the stored state.

Junctions J1 and J2 connected to a storage loop form a decision making pair (DMP) [33] or a balanced comparator [149]—a common structure in RSFQ logic, shown in Fig. 4.5. Depending upon the direction of the persistent current in the loop, junction J2 is either biased close or far from the critical current. When a clock pulse arrives, if J2 is biased close to the critical current (corresponding to logic one), J2 switches, and an SFQ pulse is produced at the output. Alternatively, if J2 is biased far from the critical current (corresponding to logic zero), J1 switches first, and an incoming SFQ clock pulse escapes through J1. The critical current of J1 and J2 is chosen to tolerate wide parameter variations while exhibiting a narrow gray zone (those parameter values where switching behavior is uncertain) [150].

An RSFQ DFF demonstrates two important elements of the RSFQ circuit design process—storage loops and decision making pairs. Most RSFQ logic gates utilize a combination of these two primitive elements [47].

4.2.2   Buffer

A Josephson junction is a two-terminal device and does not provide input-output isolation (i.e., bidirectional signal transmission) [45]. Most RSFQ circuits are reciprocal—output pulses can propagate back to the input terminals of a gate and produce errors. To prevent this behavior, a buffer structure is used.

An RSFQ buffer is schematically depicted in Fig. 4.6. It consists of two JJs, where each JJ exhibits a different critical current and $I_c(J2) < I_c(J1)$.

When an SFQ pulse arrives at terminal A (the input), J1 is biased closer to the critical current than J2. Moreover, J2 is biased in a different direction from the incoming pulse. Hence, J1 switches, producing an SFQ pulse at terminal B (the output). When an SFQ pulse arrives at terminal B (the output), J2 is biased in the direction of the incoming pulse. In this case, J2 switches first, and no SFQ pulse is produced.

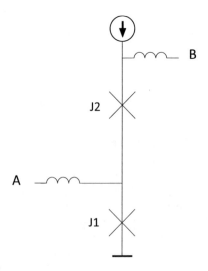

Fig. 4.6 RSFQ buffer

Although this structure can be utilized as a separate buffer gate, the buffer is frequently integrated into other logic gates to isolate the inputs from the output while reducing complexity. When included within an existing gate, only J2 is frequently necessary [54]. This junction is called an escape junction. Combined with the storage loops and balanced comparators, these structures form a common topology in RSFQ circuits.

4.2.3 Splitter

Due to the pulse-based representation of signals in RSFQ logic [52], most gates exhibit a fanout of one. An active splitter gate is therefore necessary to provide a larger fanout.

A splitter circuit is schematically depicted in Fig. 4.7. An input SFQ pulse arriving at terminal A switches junction J1. J1 is larger (exhibits a higher critical current) than J2 and J3 and therefore drives a large input current into J2 and J3, switching these junctions and producing an output pulse at both terminals B and C.

Binary splitter trees are used to provide a fanout greater than two. A four-output binary splitter tree is shown in Fig. 4.8. A clock network in RSFQ circuits contains multiple splitters; the overall number of splitters in a complex circuit can be comparable to the number of logic gates.

Three-way and larger splitter trees have also been demonstrated; however, the parameter margins of these circuits are narrower than the parameter margins of a two-output splitter. Some layout area can be saved by sharing the input and output junctions within a binary splitter tree (e.g., J_1 and J_2 in Fig. 4.8). This approach,

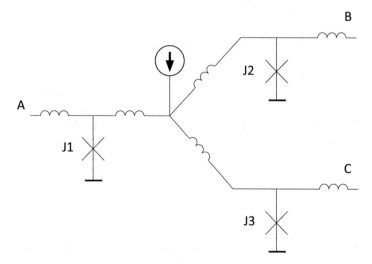

Fig. 4.7 RSFQ splitter

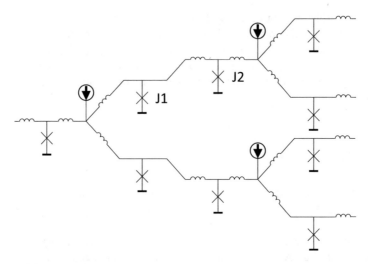

Fig. 4.8 RSFQ binary splitter tree

however, requires higher JJ critical currents for the JJs within the splitter tree. Another prospective method to save area is to use passive splitters, composed of a splitting passive stripline [151, 152]. In this approach, a passive transmission line, typically a one-to-one structure, is split into two lines and simultaneously connected to two different outputs. These structures, however, require careful impedance matching and exhibit narrow margins.

4.2.4 Confluence Buffer

An RSFQ confluence buffer (merger) provides a means to merge SFQ pulses from two different sources. This gate functions as an asynchronous OR gate [23], propagating input pulses from any of the inputs to the output [54].

The merger gate is shown in Fig. 4.9 and consists of five JJs, operating as follows. Both $J1 - J3$ and $J2 - J4$ individually form a buffer structure, as previously described. Consider an input pulse arriving at terminal A. The input pulse from A arrives at the input of buffer J1 and J3 and switches J1, which passes the SFQ pulse further along. The resulting waveform arrives at the output of buffer $J2 - J4$, switching escape junction J4 and preventing this pulse from propagating toward terminal B. Finally, this pulse switches J5 and the output is produced at terminal C. In the case of an input pulse arriving at terminal B, the operation is similar with the orientation of the buffers reversed.

In this gate, any combination of input pulses produces an output [53]. In the case of the inputs arriving at both terminals close in time, only one output is produced. This feature is exploited in a synchronous OR gate.

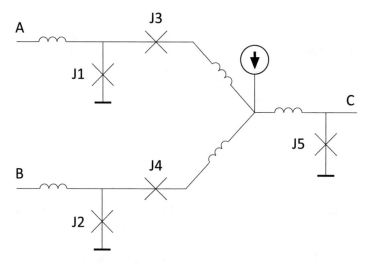

Fig. 4.9 RSFQ confluence buffer

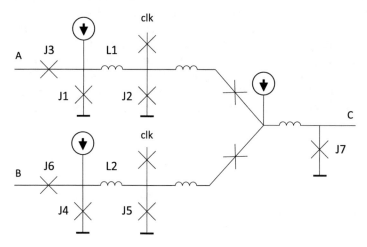

Fig. 4.10 RSFQ OR (AND) gate

4.2.5 OR and AND Gates

An RSFQ OR gate is a clocked confluence buffer. A schematic of this circuit is shown in Fig. 4.10.

The OR gate consists of two clocked storage loops (DFFs), $J1 - L1 - J2$ and $J4 - L2 - J5$, and a confluence buffer. When an SFQ pulse arrives at either of the input terminals, A or B, a flux quantum is latched into the corresponding DFF. Additional input flux quanta at the input terminals are rejected by the escape junctions, J3 and J6. When the clock pulses arrive at the clock terminals, the confluence buffer produces an output pulse if any of the inputs had arrived during the preceding clock period. The resulting gate performs a synchronous OR operation.

The RSFQ AND gate can utilize schematic similar to the OR gate [153, 154]. In the AND gate, the physical size (critical current) of junction J7 is increased to ensure that the junction can only be switched by two simultaneous input pulses (arriving within a few picosecond) [48].

4.2.6   Inverter

As compared to conventional CMOS electronics, inversion is one of the most complex basic operations in RSFQ logic. An RSFQ inverter, due to the pulse-based representation of a data signal, also must be clocked as the structure produces an output pulse without any input pulses.

An RSFQ NOT gate is schematically depicted in Fig. 4.11. The loop, $J3 - L1 - J4 - J5$, provides a storage function similar to the DFF. An SFQ input pulse switches junction J4, redirecting the bias current into J3. This state is retained until the arrival of the clock signal. When a clock pulse arrives, the logic gate switches the escape junction J2, thereby not producing any output pulses. After a small delay due to the resistor, J3 switches and resets the state to the initial state. If no input pulse arrives at terminal A, bias current is directed into the $J4 - -J5$ branch, and an incoming clock pulse initially switches J5, producing an output pulse. The escape junction J4 also switches, preventing this pulse from propagating toward the input terminal.

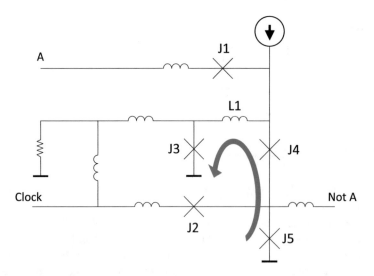

Fig. 4.11 RSFQ inverter. Arrow denotes the SFQ storage loop

4.2.7   Muller C Element

RSFQ circuits require complex clock networks. An important synchronization gate is the Muller C element [155]. This gate provides an output once both inputs arrive.

An RSFQ C element (coincidence junction) is schematically shown in Fig. 4.12. $J1 - L1 - J3$ and $J2 - L2 - J3$ each form a storage loop. Inductors, L1 and L2, and junction J3 are sufficiently large to store an input fluxon within the corresponding loop. When one of the inputs arrives, the flux quantum is stored until the second input arrives, which switches J3 and produces an output SFQ pulse.

4.2.8   SFQ-to-DC and DC-to-SFQ Converters

RSFQ circuits process single flux quantum pulses. These pulses, however, cannot exist outside of the superconductive circuits and transmission lines. To provide external access to RSFQ ICs, conversion between picosecond SFQ voltage pulses and room temperature voltage level waveforms is necessary.

A DC-to-SFQ converter is schematically shown in Fig. 4.13. The converter consists of two JJs and an inductor forming an asymmetric, two junction SQUID. If the input current increases above a specific activation current, J1 switches and produces an SFQ pulse. If the input current is lowered, J2 switches and no output is produced. This circuit is controlled by a square, sine, or sawtooth waveform, corresponding to a target bit pattern. This waveform is typically called "DC" despite the signal changing in time [23].

An SFQ-to-DC converter is schematically depicted in Fig. 4.14. The circuit consists of a storage loop, $J2 - L - J4$, similar to a DFF, with J5 and J6 attached

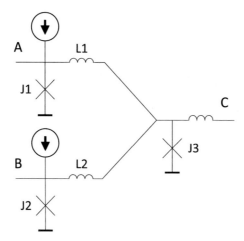

Fig. 4.12 RSFQ Muller C element

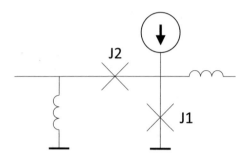

Fig. 4.13 DC-to-SFQ converter

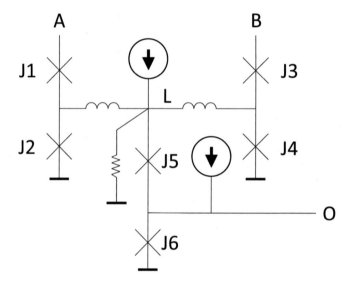

Fig. 4.14 SFQ-to-DC converter

to the center inductance of the loop [23]. An input pulse at terminal A switches J2, increasing the current in the J4 branch and $J5 − −J6$ branch. J5 and J6 are biased close to the critical current and switch into the resistive state, producing continuous voltage pulses at the output terminal O. These pulses can be averaged, resulting in a relatively high output voltage. An input pulse supplied to terminal B resets the gate into the initial state.

4.3   Bias Networks in RSFQ

The primary parameter of a Josephson junction is the critical current I_c, which affects the transition between the superconductive and voltage states of a JJ, and is directly related to the physical area of the JJ. The JJs within an RSFQ circuit are directly or indirectly biased to a specific fraction of I_c to maintain proper operation [23, 154]. Local bias current distribution within each gate is performed by an inductive network, consisting of inductors and JJs (which exhibit a nonlinear inductance), with one or two bias network connections per gate. Each gate is typically individually optimized—the inductance and critical current of the JJs are chosen to produce robust operation and small delay.

The objective of a bias distribution network within a complex RSFQ circuit is to supply a precise bias current to each gate. In the following subsections, different topologies for distributing bias current in RSFQ circuits are described, and the advantages and disadvantages of these methods are reviewed. In Sect. 4.3.1, conventional distribution of bias current in RSFQ circuits is described. In Sect. 4.3.2, energy efficient bias schemes for RSFQ circuits are reviewed.

4.3.1 Bias Distribution in RSFQ Circuits

In conventional RSFQ circuits, the bias current is distributed and regulated by a resistive tree network [23]. The current is typically supplied off-chip and transferred to the gates by superconductive wires, where each cell contains a bias resistor, as depicted in Fig. 4.15. Unlike CMOS bias networks, which exhibit some distributed resistance per length, an RSFQ bias network is lossless until the point of load [156]. Within each cell (flip flop or logic gate), the bias current is distributed by inductive current division, where the nonlinear inductance of the JJs is also considered. For example, in the JTL shown in Fig. 4.1, the bias current I_B is inductively divided between every two junctions.

The resistors within the gates dissipate significant static power, approximately 60 times greater than the dynamic power dissipated during a JJ switching event at a clock frequency of 20 Ghz. ($P_D = I_b * \Phi_0 * f_s$, ~13 nW per gate) [33]. Moreover, most of this power dissipation occurs close to the thermally sensitive superconductive elements. These issues and concerns emphasize the importance of correct and efficient distribution of bias currents within large scale SFQ circuits [49].

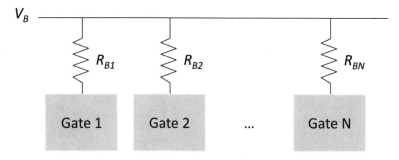

Fig. 4.15 Bias current distribution in conventional RSFQ circuits

4.3.2 Energy Efficient Bias Distribution in RSFQ Circuits

Multiple solutions have been proposed to mitigate this static power dissipation in RSFQ circuits. Some of these solutions are described in this subsection.

4.3.2.1   Resistive Bias Current Distribution with Low Bias Voltage

A straightforward method to reduce power dissipation in bias resistors is to decrease the bias voltage across these resistors [157]. This approach, however, undesirably redistributes bias current throughout the circuit.

Consider the RSFQ circuit shown in Fig. 4.16. Two logic gates (shown in the shaded boxes) are each supplied with an SFQ data signal operating at a different frequency. In the worst case, one gate receives a stream of logic "zeros" (no SFQ pulses), and the other gate receives a stream of logic "ones" (an SFQ pulse every clock period). The average voltage at the latter gate is equal to $V = \Phi_0 f_c$, where f_c is the clock frequency. The average voltage at the former gate is zero. This voltage difference across the resistance R produces a change in bias current,

$$\Delta I = \frac{\Phi_0 f_c}{R}, \tag{4.4}$$

which reduces or increases the bias current of the gates, potentially producing errors.

Fig. 4.16 Data-dependent redistribution of bias current in RSFQ circuits [156]

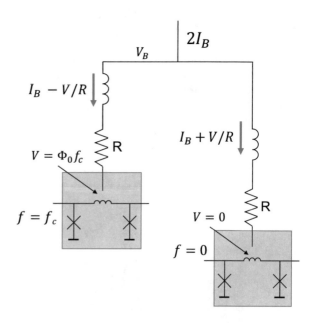

RSFQ circuits exhibit limited bias margins, typically on the order of 30%. This data-dependent change of bias current should therefore be significantly smaller than the bias margins within a circuit. This effect produces a limit on the maximum clock frequency f_m depending upon the bias voltage V_B [33],

$$f_m \sim 0.1 \frac{V_B}{\Phi_0},$$

(4.5)

where 0.1 is assumed as the maximum permissible change (10%) of the bias current.

RSFQ circuits can operate with a smaller bias voltage and resistance while dissipating lower static power [157]. This modification, however, limits the maximum clock frequency in these circuits.

4.3.2.2   Inductive Bias Current Distribution

A natural solution to the problem of static power dissipation in RSFQ circuits is to exploit the dissipationless properties of superconductive inductors. Inductive current distribution produces a system with zero static power dissipation. The issue of current redistribution, however, is exacerbated by the absence of bias resistors.

Consider the circuit shown in Fig. 4.17, which is similar to Fig. 4.16. The current between the gates depends upon the difference in phase rather than the difference in average voltage [158],

$$\Delta I = \frac{\phi_1 - \phi_0}{2L_b},$$

(4.6)

where L_b is the bias inductance. Although this inductance can be large, current redistribution becomes more significant each time the phase difference increases between the JJs within the two circuits. In the worst case, this current increases every clock cycle and eventually produces errors—additional SFQ signals in those circuits with higher bias, and a lack of expected SFQ signals in those circuits with lower bias.

The issue described here is the issue of phase balancing. To utilize inductive current distribution in RSFQ circuits, the phase of each of the bias terminals needs to be equal or sufficiently close to prevent bias current redistribution. Multiple approaches exist to achieve phase balance, some of which are described in the following subsections.

4.3.2.3   Dual-Rail SFQ

Data-dependent phase imbalance arises due to the different representation of logic "ones" and "zeros" in RSFQ circuits. A method to mitigate this issue is to represent both "ones" and "zeros" in the form of SFQ pulses, routing these signals using two separate wires, or "rails."

Fig. 4.17 Redistribution of
bias currents in inductively
biased RSFQ circuits [33]

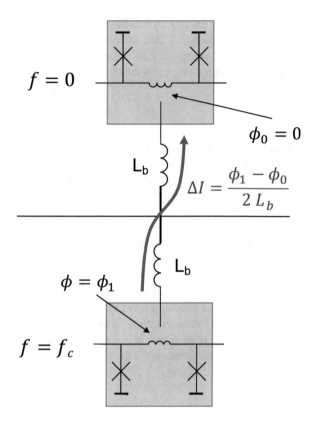

This approach is called dual-rail [158] or delay-insensitive [159] SFQ circuits. As dual-rail circuits use two rails for data signals, these circuits do not utilize conventional RSFQ primitives, as described in Sect. 4.2. The bias supply in the gates is connected to the phase balanced terminals, thereby alleviating any bias current redistribution. This inductive bias distribution approach can be used to bias gates in other logic families as long as the phase of the bias terminals is balanced. Another major advantage of these circuits is delay insensitivity due to asynchronous operation [159]. This property reduces the timing complexity of large scale SFQ systems.

Certain drawbacks, however, exist with this approach. Delay-insensitive gates with zero static power dissipation can require a significant number of additional JJs for phase balancing [158]. In addition, as every connection between gates requires two separate wires, the routing area is greatly increased.

4.3.2.4   ERSFQ

In energy efficient RSFQ (ERSFQ) [156], the dissipative resistors within the RSFQ logic gates are replaced with Josephson junctions and superconductive inductors. This modification eliminates any static power dissipation, reducing the

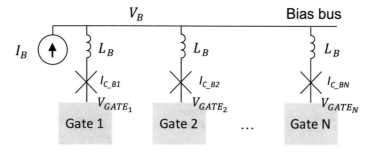

Fig. 4.18 Bias distribution in ERSFQ circuits

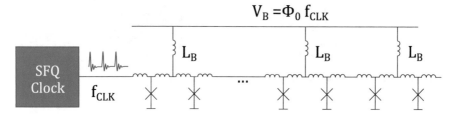

Fig. 4.19 Feeding JTL connected to an SFQ clock line acting as a voltage reference

total dissipated energy by two orders of magnitude [33]. These JJs function as current limiters—when the current passing through these bias JJs approaches the critical current I_c, the inductance of the JJ rapidly increases. If the current exceeds I_c, the bias JJ momentarily transitions into the voltage state, diverting any additional current within the bias network. Conversion between RSFQ and ERSFQ gates does not require any changes to existing cell libraries, only affecting the bias distribution elements [160]. This bias scheme is schematically depicted in Fig. 4.18.

Multiple modifications are necessary to support inductive bias distribution. Switching the bias JJs produces current fluctuations on the order of Φ_0/L_B, where L_B is the bias inductance connected in series with the bias JJ [156]. A large bias inductor L_B therefore reduces the bias current ripple, although a large inductance (hundreds of picohenries) requires significant area. This tradeoff has led to modifications in some fabrication processes, for example, the introduction of a high kinetic inductance layer [139].

The average voltage for a gate switching at a frequency f_s is $\Phi_0 * f_s$. To prevent current redistribution, the voltage on the bias bus should be higher than any gate voltage within the circuit. This constraint is achieved by connecting the bias bus to the clock line—the average voltage on the clock line is guaranteed to be equal or greater than any gate voltage, since the clock is assumed to operate at the highest frequency in a circuit. The clock line is connected to a structure called a feeding JTL (FJTL) to increase both the stability of the voltage reference and the bias margins.

A feeding JTL is schematically depicted in Fig. 4.19. A FJTL is a JTL consisting of multiple stages, where each stage is connected to the bias bus by a large inductor L_B. This JTL is typically terminated, and the output is not utilized. The FJTL

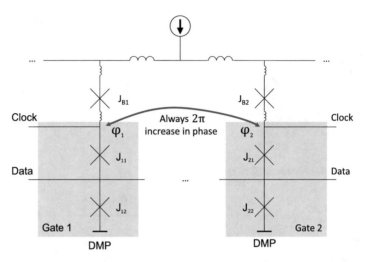

Fig. 4.20 Phase balance across different decision making pairs [167]

establishes a robust voltage reference for the bias bus of an ERSFQ circuit and improves the margin of operation by supplying additional or receiving excess bias current.

Some commonly used *ad hoc* design approaches and guidelines exist on the proper design of these bias networks [50, 161–165]. As the size and topology of the feeding JTL affect the physical area and bias currents, strict guidelines are required to integrate ERSFQ circuits into an industrial EDA design flow [51, 166]. These design guidelines are further discussed in Chap. 13.

4.3.2.5   eSFQ

A different approach to achieve a phase balance which utilizes inductive bias current distribution is used in energy efficient SFQ (eSFQ) logic [33, 167]. As previously discussed, bias current redistribution in inductively biased RSFQ circuits is due to the difference (imbalance) between the phase at different bias locations. In ERSFQ, this redistribution switches the bias JJs during circuit operation, thereby requiring large bias inductors to reduce current variations. In RSFQ circuits, however, a structure exists that maintains a constant change in phase every clock cycle—a decision making pair connected to the clock network. This feature is shown in Fig. 4.20.

In eSFQ circuits—circuits with synchronous phase compensation [33]—this feature is exploited. The bias terminals are merged with the clock terminals. The clock terminals experience a 2π change in phase every clock cycle regardless of the input data. Each DMP connected to the same clock network exhibits the same phase. By supplying the bias current through the DMP, the phases are naturally balanced, and the bias JJ only switches during the initial power-up process [33]. Conversion

Fig. 4.21 Conversion of regular RSFQ DFF into an eSFQ DFF [167]

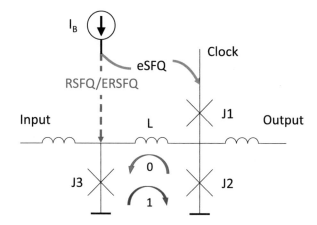

of a gate (DFF) into eSFQ is schematically shown in Fig. 4.21. Note that an eSFQ DFF initially stores a logic "one" and requires a reset.

The need for large bias inductors is eliminated in eSFQ circuits as the bias JJs do not switch during circuit operation, thereby requiring significantly less area. This approach also does not require a separate feeding JTL. The primary drawback of eSFQ is that this logic family is incompatible with existing RSFQ cell libraries. Each logic gate in eSFQ needs to be consistent with the circuit topology shown in Fig. 4.21. For many gates, these changes add to the number of JJs in a circuit, increase the dynamic power, and can also change the initial state [33]. These disadvantages, along with the addition of a high kinetic inductance layer into the fabrication process [139], lessen wider adoption of the eSFQ logic family.

4.4 Conclusions

The basic building blocks of digital RSFQ circuits are described in this chapter. Different types of transmission lines for this technology—JTL and PTL—are introduced, and the advantages and disadvantages of these interconnect structures are discussed. Certain basic structures of RSFQ circuits, flux storage loops, balanced comparators, and buffers, are described here as a framework to demonstrate the operation of more complex gates. Bias current distribution in RSFQ circuits is also discussed in this chapter. Resistive bias distribution, commonly used in RSFQ circuits, is described, and the disadvantages of this approach are reviewed. Energy efficient modifications to RSFQ logic to reduce or eliminate static power dissipation are also introduced, along with the advantages and disadvantages of these different circuit topologies. In particular, the ERSFQ bias scheme, which utilizes inductive current distribution with JJs as current regulators and feeding JTLs as voltage references, is introduced. Additional techniques to reduce and mitigate the deleterious effects of high bias currents are further discussed in Chaps. 9, 13, and 14.

Chapter 5
Synchronization

The speed of operation in a synchronous circuit is controlled by the clock distribution network. Most RSFQ circuits require a large multi-gigahertz clock network, as unlike CMOS most logic gates are clocked. AQFP circuits however utilize a multiphase AC power network for synchronization. In a self-timed asynchronous circuit, where a global clock network is absent, handshaking gates and protocols are necessary. Timing tolerances in all of these systems are extremely narrow, and an effective clock network is required for robust high performance operation. The optimal design of the clock distribution network, providing robustness against timing variations, and proper organization of the handshaking circuits are crucial to maintain correct operation of multi-gigahertz systems. Techniques providing solutions to these issues in superconductive circuits are described in this chapter.

A diagram summarizing existing RSFQ clocking approaches is shown in Fig. 5.1 [168]. RSFQ clocking is broadly divided into synchronous and asynchronous schemes. Different hybrid clocking approaches exist, combining a variety of techniques to enhance performance and robustness. Globally asynchronous, locally synchronous (GALS) methodologies, where regions and blocks synchronized by local clock signals are connected by handshaking signals, are between the synchronous and asynchronous approaches on the scale of relative synchronicity. These approaches produce a combination of inelastic pipelines, where the timing of each stage is fixed, and elastic pipelines, where the timing is self-regulating.

In this chapter, common RSFQ clocking approaches are described in the approximate order of decreasing synchronicity. In Sect. 5.1, common synchronous, hybrid, and GALS clocking approaches are reviewed. Asynchronous, self-timed, and dual-rail schemes are discussed in Sect. 5.2. Clock distribution schemes for AQFP circuits, which utilize entirely different clocking approaches, are described in Sect. 5.3.

© Springer Nature Switzerland AG 2022
G. Krylov, E. G. Friedman, *Single Flux Quantum Integrated Circuit Design*,
https://doi.org/10.1007/978-3-030-76885-0_5

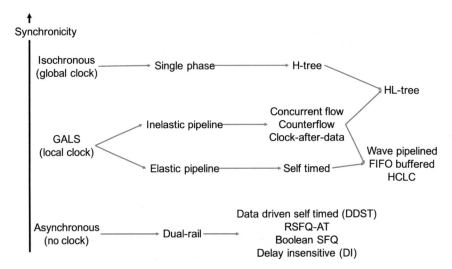

Fig. 5.1 Clocking approaches for RSFQ circuits [168]

5.1 Synchronous RSFQ Circuits

The most basic clock distribution network to synchronize an RSFQ circuit is an H-tree network [169]. This topology is schematically depicted in Fig. 5.2a. In this approach, the clock network is equipotential [170] and designed to exhibit zero clock skew between cells [171]. In RSFQ, an H-tree clock distribution network is typically composed of a binary tree of splitter gates. The length of the interconnect segments is balanced to increase symmetry and decrease variations in the clock arrival times. A major advantage of this approach is the straightforward design process [170]. Disadvantages are the poor tolerance to timing variations, and large and symmetric clock networks require significant physical area and bias current [49, 51]. No large circuits exclusively utilizing a zero skew H-tree for synchronization have to date been experimentally demonstrated. In [172], a zero skew H-tree clock network is used to synchronize an 8-bit processor; the circuit was not fabricated. In [173], an automated H-tree clock network synthesis algorithm is proposed, where the splitter delays and placement blockages are considered.

5.1.1 Common RSFQ Clocking Schemes

Several synchronization approaches have been proposed for RSFQ circuits [23]. In a concurrent clock distribution network, clock and data propagate in the same direction through the logic path. The concurrent clocking approach is illustrated in Fig. 5.2b. While the data pulse propagates toward the next cell along the logic

Fig. 5.2 Synchronous
clocking schemes for RSFQ
circuits, (**a**) H-tree, (**b**)
concurrent flow, and (**c**)
counterflow. The letter "S"
denotes an RSFQ splitter

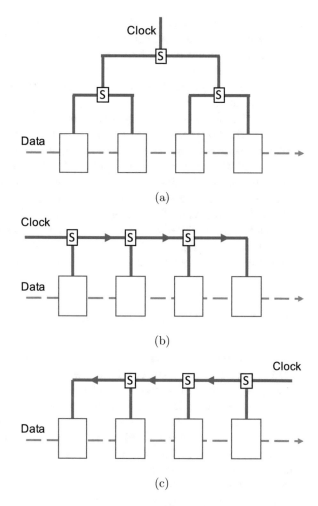

path, it lags behind the clock pulse before the data signal arrives at the next cell. The clock pulse produces an output, resetting the state of the next logic cell and preparing the cell for the arrival of the incoming input signal. Clock skew between concurrently clocked gates is negative [170, 174], which means the clock signal arrives at the initial gate before it arrives at the final gate between two sequentially adjacent gates. This clocking scheme increases the maximum clock frequency of a circuit, while degrading timing tolerances.

   A counterflow clock distribution network is similar to the concurrent clock network, but the clock signal propagates in the opposite direction with respect to the data [23]. The counterflow clocking approach is shown in Fig. 5.2c. When a gate receives a clock signal, the next gate in the logic path has already received the clock signal during the same clock period. The output signal will likely arrive at the next gate when the state of that gate has already been reset. Clock skew in the counterflow scheme is positive [170, 174], which means the clock signal arrives at the initial gate after it arrives at the final gate between two sequentially adjacent gates. Counterflow clocking greatly increases the timing tolerances of a circuit [175]. The performance of the resulting system is, however, reduced as compared to the H-tree (zero skew) and concurrent (negative skew) approaches due to the positive clock skew (which adds to the path delay).

A variation of the concurrent scheme is the clock-follow-data scheme [171]. In this approach, the (negative) clock skew is further increased as compared to the concurrent clock network. A clock pulse arrives at the next cell in the logic path after the arrival of the corresponding data pulse. The data propagates through the entire logic path based on the single clock pulse at the beginning of the logic path. Logic paths clocked using the clock-follow-data scheme are asynchronous with respect to the rest of the circuit. The timing of these paths depends upon the local clock skew, purposely introduced by the delay elements, and is not dependent on the clock period. This approach produces circuits that are highly robust to timing variations, but require significantly more area and exhibit reduced throughput.

5.1.2 Hybrid Clocking Approaches

Different combinations of these clocking approaches are used in practical RSFQ circuits. In general, the concurrent clocking scheme exhibits significant performance advantages as compared to the other schemes. This advantage is primarily due to the highly pipelined architecture of RSFQ circuits, where each logic gate is clocked and the data path between any two sequentially adjacent synchronous gates exhibits a small difference in delay [176]. In [177], a 32-bit Kogge-Stone adder (KSA) operates at a clock frequency over 50 Ghz. In this KSA, the concurrent flow clocking scheme is between the pipeline stages, while each pipeline stage receives a clock signal distributed by a binary H-tree network. In [178], the HL-tree topology is proposed. In this topology, the global H-tree network is connected to concurrently clocked segments (or leaves). The cells are grouped by the logic level, enabling an abutted connection of clock distribution elements, thereby reducing the area of the circuit [179].

Clocking a logic path containing feedback loops is a significant challenge. The sum of all clock skews across a loop must be zero [180, 181]. Both the concurrent approach, where the net clock skew is negative, and the counterflow approach, where the net clock skew is positive, are therefore not directly applicable for clocking a loop structure with feedback. Hybrid approaches combining both clocking schemes have been proposed. The simplest structure composed of a logic loop is a circular shift register (CSR), in which no operations are performed while the data propagates around the loop. Hybrid approaches are frequently evaluated using CSR structures. In one of these hybrid clocking approaches, a CSR is divided into two parts, one part is clocked concurrently, while the other part uses a counterflow scheme [182]. This approach is expanded in [183] for larger CSR structures. The homogeneous cloverleaves clocking (HCLC) approach, proposed in [184], utilizes pairs of leaves with concurrent and counterflow clocking and Muller C elements for synchronization. In [185], hierarchical chains of HCLC structures are proposed to extend this approach to large scale circuits.

5.1.3 GALS

Large scale ICs frequently do not utilize a single global clock signal common to all cells. Different regions are synchronized by different clock signals (multiple clock domains), possibly with different distribution topologies and frequencies, and the synchronicity of the resulting system is reduced, as shown in Fig. 5.2. These regions are connected by specific synchronization circuits, necessary to cross these clock domains. These approaches can be broadly defined as globally asynchronous, locally synchronous (GALS) clocking [52]. The GALS technique is widely used in modern CMOS circuits to reduce the complexity of the clock distribution network [186]. Although modern RSFQ circuits exhibit significantly lower complexity as compared to CMOS circuits, due to the aforementioned timing challenges of RSFQ logic, GALS approaches are also common in RSFQ.

The majority of the largest fabricated and tested RSFQ circuits utilizes a GALS synchronization methodology. The FLUX-1 processor [187] is based on a localized clocking technique, similar to GALS. This 16-bit processor operates at a frequency of 20 Ghz. Specific regions within this processor utilize an H-tree or a concurrent flow clock scheme, while the clock skew between the regions is externally controlled [188]. The 8-bit bit-serial RSFQ processor CORE e4 has been demonstrated to operate at a local clock frequency of 80 Ghz, while the system clock frequency is 2 Ghz [134]. This processor also utilizes fast local clock signals for bit-serial operation.

5.1.4 Dynamic SFQ

Dynamic SFQ (DSFQ) [189] is a recently introduced type of RSFQ logic, where the state of a gate is temporarily stored, and the gate self-resets to the initial state after a period of time. The storage and reset time are adjustable, producing a tradeoff between the input skew tolerance and performance [48]. This capability enables asynchronous operation of logic gates, significantly reducing the size and energy requirements of complex clock networks in large scale RSFQ circuits. Although dynamic flip flops can also be constructed, it is preferable to use conventional, clocked flip flops to separate the pipeline stages. In this way, dynamic SFQ circuits are more similar to CMOS than standard RSFQ circuits and can be separated into sequential and combinatorial logic [189], enabling the use of relevant CMOS EDA techniques. A more complete discussion of DSFQ logic is provided in Chap. 11.

5.2 Asynchronous RSFQ Circuits

Clocked logic gates is a major challenge of synchronous RSFQ circuits. Apart from
the timing issues, N cells (clock sinks) require at least $N - 1$ binary splitter gates
to distribute the clock signals. With additional space required for routing, this issue
translates to significant area occupied by the clock distribution network. Multiple
asynchronous approaches have been proposed to reduce the overhead of clocking
RSFQ circuits [190].

5.2.1 Handshaking Protocols

To reuse conventional RSFQ gates while enabling asynchronous operation, a
handshaking mechanism is necessary. Multiple handshaking circuits have been
proposed [23, 168]. These circuits, in general, exhibit lower area overhead in RSFQ
technology as compared to CMOS, further incentivizing the use of asynchronous
schemes. A typical handshaking mechanism utilizing Muller C elements (coincidence junctions) is shown in Fig. 5.3. A C element, described in Sect. 4.2.7,
produces an output signal as soon as both inputs arrive. Each stage within a
path receives a request signal (REQ) from the previous stage in the path and an
acknowledgement signal (ACK) from the next stage in the path. As soon as both
signals are received, the C element synchronizes the stage and sends handshaking
signals to the neighboring stages.

When a handshaking approach is applied to a shift register, a FIFO buffer is
produced [168]. A FIFO is the least complex example of an elastic pipeline, while
a shift register is the least complex example of an inelastic pipeline [191]. In an
inelastic pipeline, the processing time and therefore the clock delay of each stage
is fixed. An elastic pipeline automatically adapts to any changes in the processing
time. The timing of each stage only depends upon the timing of the neighboring
stages.

In a self-timed [192] or resynchronized [171] system, both counterflow and
concurrent clock paths are utilized. A C element produces a clock pulse once

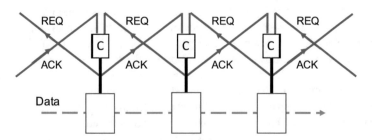

Fig. 5.3 Handshaking mechanism based on C elements

both clock inputs have arrived. This scheme significantly reduces the clock skew and enables self-timed operation. The primary disadvantage of this approach is the significant additional area required for the synchronization elements. A source for the clock pulse train is necessary for this approach, distinguishing resynchronized systems from other asynchronous styles.

In a FIFO buffered clocking approach [168], a FIFO buffer connects synchronous blocks with different clock phases. The FIFO buffer saves the data sent by one block until another block can receive the data. If the FIFO buffer is sufficiently large, this approach can reduce or eliminate timing uncertainties between blocks.

In [193], a wave-pipelined clocking approach [194] is used within an 8-bit ALU based on a KSA. The computation in each stage starts when both operands are present. The clock-follow-data approach resets the cells to the initial state between operations. This ALU has been experimentally demonstrated to operate at 20 Ghz [193]. A distinctive feature of this ALU is a wide data path, as opposed to the bit-serial architecture common in earlier RSFQ processors [195]. A similar approach has been used for a 16-bit parallel adder [196].

5.2.2 Dual-Rail Logic

Dual-rail logic [197, 198] utilizes a differential signaling protocol, different from conventional RSFQ circuits. In this type of logic, all connections between gates are composed of two wires (rails), as opposed to one wire in conventional RSFQ. An SFQ pulse on one wire represents a logic "one," while an SFQ pulse on another wire represents a logic "zero." This approach alleviates the need for a clock signal, as the gates can operate as soon as an input signal is received on any of the wires. A dual-rail gate is naturally larger than an RSFQ gate, requiring twice as much resources for signal routing. Significant area savings, however, is achieved from the asynchronous operation. Another significant advantage is the trivial inversion operation (the signal wires are swapped), which is costly in standard RSFQ. Boolean SFQ (BSFQ) [199] is a similar dual-rail approach that utilizes set and reset wires for each signal.

Another dual-rail approach is RSFQ-AT logic [200]. In RSFQ-AT, a clock signal is attached to each data signal; global synchronization is therefore not necessary. RSFQ-AT cells are typically composed of regular RSFQ circuits and timing circuitry. The timing circuits synchronize the overall circuit and generate the output clock signals with appropriate delays. In this approach, the complexity of the clocking system is constrained within each logic cell. The clock signal propagates together with the data signal, and the cells are designed to satisfy the correct timing constraints.

In data-driven self-timed (DDST) RSFQ circuits, two wires similarly carry complementary data pulses [201]. In this approach, the logic gates are the same as conventionally clocked RSFQ circuits. The clock signal is recovered at each logic gate by merging complementary data pulses. The output signal produced by the gate is converted into the complementary form. The TIPPY3 RSFQ processor [202] utilizes this DDST clocking technique. This 8-bit bit-serial RSFQ processor has been experimentally demonstrated to operate at a local clock frequency of 10 GHz. In the fully asynchronous bit-serial processor SCRAM2 [203], handshaking and DDST approaches are both used. Dual-rail gates are different from conventional RSFQ gates; however, the same circuit primitives—splitters, mergers, storage loops—are used as well as a similar set of logic gates.

Delay-insensitive (DI) logic is based on a different set of primitives, first proposed in [204]. DI gates comprise a universal logic set. Among these primitive gates are "join," "fork," and "tria," which in RSFQ technology exhibit similar complexity as conventional gates [159]. Similar to dual-rail gates, DI gates do not require synchronization. Another important advantage of DI logic is compatibility with inductive bias distribution networks [33]. Zero static power is therefore dissipated without requiring additional bias techniques [50, 158]. Significant advantages of DI circuits, such as lower latency and smaller area as compared to dual-rail approaches, have been demonstrated [190]. The primary disadvantage of this logic is shared with other dual-rail approaches—significant additional area is required to route the interconnect as compared to single-rail logic families. In addition, circuit modifications to achieve phase balance can increase the area of the gates.

5.3   AQFP Circuits

Clock distribution in AQFP circuits is significantly different than in RSFQ circuits. AQFP circuits are synchronized and biased by an AC signal. This feature provides greater compatibility with CMOS clocking techniques, as opposed to pulse-based RSFQ clock networks.

To enable forward data movement in an AQFP pipeline, the AC clock is necessarily multiphase. For every logic level, the next level is synchronized by the following clock phase. In modern AQFP circuits, a four-phase clock signal is common [118], supplied by two AC current sources [206], each phase shifted by $90°$. A common technique to distribute this clock signal to a long data path is to utilize a meander topology [205]. This topology is schematically shown in Fig. 5.4. The first four logic levels are clocked by increasingly shifted AC signals before the first clock phase is reused. The primary disadvantage of this approach is that the round trip length of the clock lines must be much smaller than the wavelength of the clock signal [205] to avoid large clock skew between the rows.

Alternative clock topologies have been proposed for AQFP circuits. In [205], a block-based clock topology is described. Large blocks in AQFP produce a long round trip time for the clock signal propagating within the meander topology, increasing the clock skew between rows. The clock frequency is therefore often lowered to avoid errors. In the block-based approach, the circuit is divided into blocks, and each block is separately synchronized using a power divider. This topology divides the larger blocks with large clock skew between the rows into smaller blocks, thereby increasing the maximum clock frequency. Delay line clocking is described in [206]. In this approach, a single clock source and phase are utilized. The phase of the clock is shifted between the cells by the delay lines. In

Fig. 5.4 Meander topology for the clock distribution network in AQFP circuits [205]

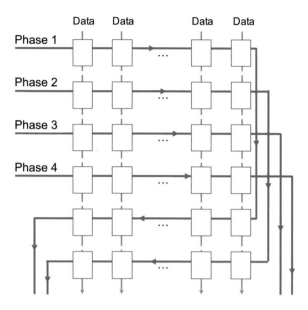

[207], a microwave power divider generates different clock phases. This approach produces a smaller delay between adjacent logic stages, increasing the performance of the overall system.

In AQFP, it is similarly challenging to provide robust synchronization for logic loops (feedback). In [208], a hybrid clock distribution topology is proposed. The circuit is divided into two blocks—concurrent and counterflow. The data signal can therefore propagate in both directions relative to the clock signal.

5.4 Conclusions

Robust synchronization techniques are essential for the correct operation of superconductive electronic circuits, particularly important due to the multi-gigahertz operation. In this chapter, synchronization of superconductive circuits is discussed. Some common, fully synchronous approaches to RSFQ clocking are described. These approaches demonstrate important RSFQ synchronization concepts and are applicable and widely used in smaller circuits and blocks. Approaches with local clock signals, similar to the GALS topology in conventional circuits, as well as other hybrid approaches, are also reviewed. These mixed clocking techniques enable sufficient robustness and performance of the clock network in large scale circuits. Asynchronous, self-timed, and dual-rail approaches, providing enhanced tolerance to timing variations, are also discussed. Examples of usage—processors and large scale circuits based on these approaches—are reviewed. AQFP circuits require an entirely different clocking mechanism as compared to RSFQ—AC signals rather than SFQ pulse trains. Common AQFP clocking approaches, as well as novel developments in AQFP clocking, are also described in this chapter.

Chapter 6
Superconductive IC Manufacturing

The key to large scale integration is advanced and robust manufacturing, enabling high density devices and interconnect while maintaining reasonable control of the process parameters. While the basic principles of integrated circuit fabrication apply to superconductive circuits, significant differences exist in materials, tolerances, and other factors.

In this chapter, the different steps, materials, and challenges that exist in fabricating modern superconductive integrated circuits are reviewed. In Sect. 6.1, the standard steps, materials, and challenges of a fabrication process for superconductive ICs are described. In Sect. 6.2, relatively novel features and current and future research directions for SCE IC manufacturing are discussed. The chapter is summarized in Sect. 6.3.

6.1 Superconductive IC Fabrication Process

The standard fabrication steps for modern superconductive integrated circuits are described in this section. Several state-of-the-art fabrication facilities currently exist for superconductive circuits—at MIT Lincoln Laboratory (LL) [209], the National Institute of Advanced Industrial Science and Technology (AIST) in Japan [210], and Hypres Inc. [211]. Several other facilities are capable of manufacturing superconductive circuits at lower complexity and integration scale. The available published information on these manufacturing processes is briefly summarized here to introduce the essential steps of a superconductive IC fabrication capability. While many steps are similar to the fabrication of conventional semiconductor circuits, several notable differences exist and are emphasized in this discussion.

The primary characteristics of a superconductive IC fabrication process are the critical current density of the Josephson junctions, the number of metal layers, and

© Springer Nature Switzerland AG 2022
G. Krylov, E. G. Friedman, *Single Flux Quantum Integrated Circuit Design*,
https://doi.org/10.1007/978-3-030-76885-0_6

the minimum feature size. Improving these characteristics is the primary driver for scaling superconductive digital circuits.

As discussed in Chap. 4, the critical current I_C of each JJ is one of the primary design parameters for SFQ circuits, which deeply affects the circuit function and parameter margins [45]. SFQ gates and flip flops are designed with a specific I_C for each JJ [23]. The critical current of a JJ is proportional to the area of the JJ and the critical current density J_C of the fabrication process [212]. The critical current density J_C therefore determines the physical area of the JJs for a given process. A higher critical current density reduces the size of the JJs, potentially increasing the circuit density.

In superconductive fabrication, the metal layers are used both as interconnect and within the active devices. A large number of metal layers are essential for increasing IC complexity. The importance of this metric is further emphasized in SCE since the passive transmission lines require two or three metal layers, as discussed in Chap. 3. Additional metal layers are also necessary for distributing the bias and ground planes [51] and to provide shielding where necessary. A large number of metal layers also reduces wiring congestion and overall wire length.

The minimum feature size determines the size of the smallest features within a circuit. In superconductive electronics fabrication, the JJs typically do not feature the smallest dimensions. This minimum size typically corresponds to the smallest width in one of the metal layers [78]. Scaling the minimum feature size reduces the area required by the gates and wiring, enabling higher complexity circuits.

Advanced fabrication processes currently exhibit a critical current density of over 100 $\mu A/\mu m^2$ (commonly written as 10 kA/cm^2), 8 to 10 niobium layers, and a minimum feature size ranging from 250 to 800 nm. These characteristics are similar to the complexity of CMOS circuits during the mid to late 1990s [78]. The primary objectives in improving manufacturing technology, apart from enhancing these characteristics, are to increase IC yield while lowering parameter variations on-chip, within different ICs on one wafer, and across different wafers.

Integrated circuits consist of a stack of materials, composed of metals, insulators, and certain layers necessary to create the active devices. An example of this stack is shown in Fig. 6.1. These layers are typically placed on a silicon wafer substrate, which is a slice of a bulk silicon crystal. Different materials are deposited and patterned to produce a target geometry. Fabrication of a thin film integrated circuit is a multi-step process, where different steps are performed for each layer, as shown in Fig. 6.2. These steps—deposition and patterning—are briefly described in, respectively, Sects. 6.1.1 and 6.1.2. The process of manufacturing a Josephson junction is outlined in Sect. 6.1.3.

6.1.1 Material Deposition

In superconductive circuits, the metal layers are not only used for interconnect but also as resistors and as contacts for the Josephson junctions. Thin metal

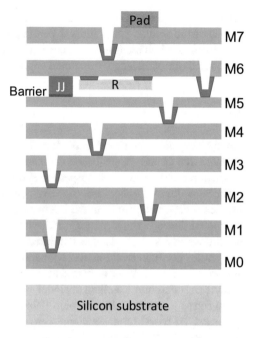

Fig. 6.1 Simplified layer stack in modern SCE circuits based on the MIT LL SFQ5ee process. The white areas are filled with inter-layer dielectric

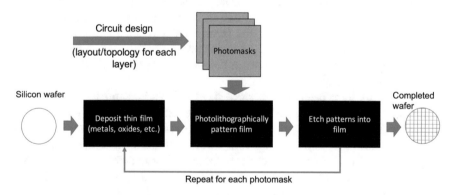

Fig. 6.2 Simplified flowchart highlighting the primary steps of an SCE fabrication process [213]

films are formed in a sputter deposition (sputtering) process—a type of physical vapor deposition (PVD) [139]. The dielectric layers are typically deposited by low temperature chemical vapor deposition (CVD) [214]. In a PVD process, the target material to be deposited is bombarded by high energy plasma, ejecting particles from the target into the chamber. This ejected material is deposited onto the substrate (wafer) through different means [139]. In the CVD process, precursor materials are introduced into the chamber in a gaseous form and interact with the substrate. The purity and characteristics of the resulting thin films depend upon the methods and materials used in this process [215]. In the following subsections, common materials used in SCE fabrication are described along with related applications.

6.1.1.1 Metal Layers

Different materials have historically been used to produce Josephson junction ICs. Early JJs and superconductive circuits were based on lead (Pb) and lead alloys [216]. Lead-based JJs exhibit good electrical properties and can be cooled with liquid helium [217]. The behavior of these circuits, however, degrades from thermal cycling [218]—a significant drawback, limiting the lifespan of cryogenic circuits, which are repeatedly cooled. Despite the significant material development efforts targeting various Pb alloy-based JJs, these devices exhibited a 5% failure rate after only a hundred thermal cycles [219]. In addition, these junctions were mechanically fragile and degraded from exposure to water or humid air [217].

In contrast, aluminum (Al) exhibits excellent material properties for thin film JJ fabrication [217] and forms a robust and well controllable oxide, which can be used as a tunneling barrier. Al-based tunnel junctions are well known and have been in use since the 1960s [220]. The primary disadvantage of Al-based JJs, however, is the low critical temperature of aluminum—1.2 K. This temperature is below the temperature of liquid helium (4.2 K). Al-based circuits therefore require significantly more expensive cooling techniques and equipment [217]. Aluminum-based Josephson junctions are widely used in circuits targeting mK operation, such as qubits [221] and related control and measurement systems [222]. In Nb-based fabrication, aluminum is an essential component—a thin layer of aluminum oxide is used as the tunneling barrier between the niobium layers within the JJ, as later described in Sect. 6.1.3.

Niobium JJs have been studied since the 1960s and emerged as the primary material for JJ fabrication in the mid to late 1980s. Although Nb can be oxidized to form a Nb_2O_5 tunneling barrier, the oxidation process is complex, and the resulting barrier contains many layers of different oxides with different characteristics and thicknesses [217]. This property results in frequent, severe defects in the tunneling barrier; the electrical properties of these junctions significantly vary within the same facility and process. A significant breakthrough in the fabrication of Nb-based JJs is the introduction of aluminum oxide as a tunneling barrier [223]. Fabrication of this oxide is well understood, and aluminum oxide exhibits a highly controllable thickness [216]. Nb also exhibits beneficial properties for the IC design process—a relatively short penetration depth and long coherence length [214]. Modern fabrication processes for superconductive digital electronics targeting liquid helium operation (4.2 K) primarily utilize niobium for the superconductive layers and JJs.

Niobium nitride (NbN) is another metal commonly used in superconductive circuit fabrication. NbN exhibits a significantly higher critical temperature ($\sim$ 16.5 K) and can therefore be used to produce circuits capable of operating at a higher temperature (over 10 K). This property reduces by more than twofold the energy required for refrigeration [214]. An important advantage of NbN for high frequency applications is the high superconductive energy gap frequency (1.2 Thz [224] as compared to 650 Ghz for niobium [225]). Thin NbN films also exhibit efficient heat dissipation [214]. Due to these properties, NbN fabrication is commonly used for superconductive detectors and other small scale, high frequency applications [226, 227].

NbN-based JJs are however poorly compatible with the aluminum oxide tunneling barrier used in Nb JJs [214]. Other tunneling barrier materials are used (e.g., MgO, AlN); maintaining uniformity of the resulting JJ critical current density is however difficult [215, 216]. The larger penetration depth of NbN also requires thicker metal layers [214]. NbN fabrication technology is less mature than Nb-based technology and is therefore not commonly used in medium and large scale digital SCE circuits [228].

6.1.1.2 Resistors

Resistors in superconductive circuits are primarily used for two purposes—as part of the RSFQ bias distribution network and to shunt the JJs. Other uses include PTL impedance matching [107, 146] and circuit design [48]. Controlling variations in the sheet resistance across a wafer is therefore essential to maintain correct circuit operation. Thin film resistors in superconductive circuits are typically fabricated from normal (non-superconductive) metals. For example, although molybdenum exhibits superconductive properties at a temperature below 1 K [229], in SCE targeting liquid helium operation, molybdenum is used as one of the materials for the integrated resistors.

While higher sheet resistance materials can produce shorter resistors, the resistors require contact vias to connect to other components. The minimum distance between these vias is however limited. For materials with high resistivity, this issue can increase the width and area of the resistors [78]. Nitrogen compounds, such as MoN_x and NbN_x, are beneficial in this case. For these materials, the resistivity can be modified over a wide range by changing the level of nitrogen within the compound [228].

6.1.1.3 Inter-layer Dielectric

In a multilayer integrated circuit, different metal layers are separated by an inter-layer dielectric (ILD) [230]. The quality of these layers affects the inductance, noise coupling, and isolation characteristics [214]. Similar to conventional semiconductor circuits, silicon dioxide (SiO_2) is the most common ILD material in superconductive circuits [228]. Other materials, such SiO and Nb_2O_5, are also used [214].

6.1.2 Patterning

High circuit density is primarily enabled by the capability to produce fine patterns within the deposited layers. These features are created using a photolithographic process.

In this process, a wafer is coated in a layer of photoresist material—a light-sensitive polymer [214]. This material is exposed to laser light at specific locations. Photoresist in these exposed locations is developed (chemically removed), while the remaining photoresist is reinforced. The resulting wafer is etched—the exposed material not protected by photoresist is removed using chemical solutions and/or plasma (i.e., reactive ion etching). Different solutions are used to etch metals and dielectrics. The remaining photoresist is removed once the layer is finalized. This process is repeated in the following layers.

The complexity and cost of the photolithographic process depend upon the target resolution (minimum feature size). In state-of-the-art submicrometer SCE fabrication, deep ultraviolet (UV) lithography is used [230]. The many significant advances developed for fabricating semiconductor ICs are exploited, as the patterning process is similar for SCE.

6.1.3 Josephson Junctions

Niobium-based JJ fabrication was introduced in the 1980s [223] and is currently the most widespread JJ fabrication process for SCE [214, 228]. A trilayer Nb/Al/AlO$_x$/Nb process is comprised of a niobium superconductive base and counter electrode layers, an aluminum layer and an aluminum oxide insulating layer. A cross section of this junction is schematically shown in Fig. 6.3. The primary steps of this process are outlined as follows.

The bottom metal and dielectric layers are initially deposited, patterned, and etched. The bottom Nb layer within the JJ (the base electrode) is deposited and covered by a thin (a few nanometers) layer of aluminum. Although aluminum is not generally superconductive at 4.2 K, this layer exhibits certain superconductive

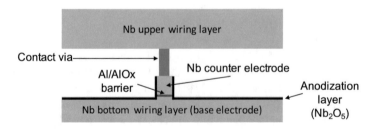

Fig. 6.3 Cross section of a thin film Nb/Al/AlO$_x$/Nb Josephson junction. The space in the middle is filled with silicon dioxide

properties. At the interface of a superconductive and normal material, Cooper pairs can penetrate to a significant depth into the normal material. This phenomenon is known as the proximity effect [231]. This layer is oxidized to form a layer of aluminum oxide with a thickness on the order of one nanometer—a few monolayers of material [232]. The critical current density of the JJs exhibits an exponential dependence on this thickness [228]. The uniformity of this oxide layer is therefore essential. By adjusting this step, the critical current density can be changed by several orders of magnitude [153, 154, 214].

The top niobium layer of the JJ (the counter electrode) is deposited on top of this oxide layer and etched where necessary to produce JJs with a specific area. Thin film JJs can be circular or rectangular. Rectangular junctions are easier to design and produce at coarser fabrication facilities; circular junctions, while less susceptible to process variations, require finer and more expensive patterning [212]. The resulting structures are anodized to protect the junctions during the following fabrication steps. The base electrode is patterned and etched to separate the junctions and to produce any necessary connections.

6.2   Manufacturing Features and Challenges

Novel fabrication techniques and features can drastically increase the integration scale of superconductive circuits. In this section, the distinctive features of modern SCE fabrication technologies are described. Some of these techniques are well established in both semiconductor and superconductor fabrication, but only recently became beneficial or feasible for SCE manufacturing.

6.2.1   Planarization

Older fabrication processes for superconductive ICs were not planarized, requiring a thick inter-layer dielectric. In a planarized process, the dielectric thickness can be shrunk, reducing the inductance per unit length. This property is beneficial for SCE digital circuits, as a thinner dielectric lowers the impedance of the transmission lines while providing higher efficiency magnetic coupling [215]. In addition, photolithography in modern advanced manufacturing processes uses a short-wavelength UV light [230]. The depth of focus—the distance over which the image is sufficiently sharp—is small in this case, requiring the preceding layers to be extremely flat.

To planarize an uneven layer within the inter-layer dielectric, a chemical mechanical polishing (CMP) process is used, in which corrosive chemicals are combined with mechanical polishing by a rotating plate to gradually strip the material. Scanning electron microscopy and ellipsometer measurements are performed to ensure the uniformity of the thicknesses of the layers [230]. Most modern SCE fabrication processes are planarized, while conventional non-planarized processes are used in low complexity, low cost applications.

6.2.2 High Kinetic Inductance Layer

One of the primary limiting factors to scaling SCE circuits is the ubiquity of inductors. These inductances affect the function, operation, and parameter margins of all of the logic gates and flip flops. SQUID loops, which store and manipulate the internal state of the logic cells, require a relatively large inductance, which translates to significant area [78]. ERSFQ bias schemes, previously described in Sect. 4.3.2.4, benefit from an extremely large inductance (hundreds of picohenries) [49, 166]. A large bias inductance reduces the transient variations of the gate bias current due to switching the bias JJs [50]. These factors emphasize the importance of reducing the size of the inductors.

The kinetic inductance, previously discussed in Sect. 2.3.10, enables a large inductance in a compact area. Several high kinetic inductance materials exist (NbN_x, TiN_x, MoN_x) and are widely used in superconductive detectors [47, 83]. Integrating these materials into a multilayer fabrication process targeting digital electronics while maintaining small variations is however challenging [139].

6.2.3 Self-Shunted Junctions

RSFQ circuits utilize non-hysteretic (critically damped or overdamped) JJs, as previously discussed in Chaps. 2 and 3. In most fabrication technologies, this damping is achieved by connecting a shunt resistor in parallel with a JJ. The area required by this shunt resistor and related connections (vias) frequently exceeds the area occupied by the JJ. Scaling the critical current density, thereby reducing the JJ area, will therefore not have a significant effect on overall circuit density.

Josephson junctions can become self-shunted (non-hysteretic) under specific fabrication conditions [233]. This effect is due to a large subgap conductance (leakage current), which is typically caused by defects in the tunneling barrier. Self-shunting is typically (but not necessarily [121]) observed in JJs with a high critical current density. Alternatively, different materials, rather than aluminum oxide, can be used to produce a tunneling barrier, e.g., doped silicon [121], or artificial defects can be introduced into the barrier. Self-shunted JJs can significantly decrease the area required by the JJs by eliminating the external shunt resistor, thereby greatly increasing the circuit density.

6.2.4 3-D Integration

Three-dimensional (3-D) integration is essential for continuing increases in the complexity of modern CMOS systems [234]. Three-dimensional memory circuits, such as HBM [235] or 3-D NAND flash [236], are currently widely used in consumer electronics, while significant research efforts are directed at novel 3-D computing architectures and heterogeneous integration [237]. In conventional semiconductor-based circuits, the active devices (transistors) utilize a silicon substrate with separate regions for the different doping types (NMOS and PMOS). 3-D integration introduces one or multiple additional device layers on top of the existing layer stack, drastically increasing the available transistor density.

The primary advantage of superconductive circuits for 3-D integration is that the substrate layer is not used by the active devices. The JJs are fabricated between the metal layers and can, in principle, be located anywhere within the metal stack. It is therefore possible to introduce additional JJ layers without adding substrate layers or through substrate vias (TSVs) [234]. Adding more active JJ layers to an SCE fabrication technology is an active area of research [210, 238]. Although the process steps for these additional layers are similar, maintaining uniformity in the process parameters is a challenging issue.

6.3 Conclusions

Manufacturing integrated circuits is a complex and intricate process, both for semiconductor and superconductive electronics. Despite the minimum feature size and number of layers in modern SCE technology being less deeply scaled as compared to semiconductor technologies, numerous additional challenges exist in manufacturing superconductive electronics. The materials used during the different fabrication steps interact in complex mechanical, chemical, and electrical ways, requiring adjustments to the manufacturing process. These issues are exacerbated by the high sensitivity of superconductive circuits to process variations.

A robust and high complexity manufacturing process is essential for large scale integration of superconductive electronics. These advanced manufacturing capabilities, however, need to be supported by specialized circuit design techniques, cryogenic test methodologies [54], and EDA algorithms. Only by combining these diverse efforts can VLSI complexity be achieved for superconductive circuits.

Chapter 7
EDA for Superconductive Electronics

Electronic design automation (EDA) is essential for the computer-aided design (CAD) of large scale systems [44]. Integrated circuit design flows can be separated into two categories. In a full custom design process, all (or most) circuit elements and devices are individually customized to produce the necessary operation. Early superconductive circuits primarily utilized a fully custom design methodology [239]. Although a circuit designed in this manner can be highly efficient, the design process is prohibitively complex and time consuming for large scale circuits. In a semi-custom design process, specific components within a system are individually designed and characterized, and the overall system is composed of multiple versions of these precharacterized components (or cells). In particular, a common approach to this semi-custom design process is the standard cell design process [240]. In this approach, commonly used elements are designed as standard cells, typically with a common height, which are reused to produce different circuits. Due to the common height, these cells are placed in standard height rows with routing between the rows, as illustrated in Fig. 7.1.

Superconductive electronics has recently reached the complexity for which a semi-custom approach is preferable in terms of computational effort, design time, and cost, as compared to a custom design process. A semi-custom design flow enables the automation of many stages of the design process.

Common steps of a typical EDA flow are schematically shown in Fig. 7.2. Specific processes, methodologies, algorithms, and tools exist for different layers of abstraction—the register transfer, logic gate, circuit, layout, and device layers. At each layer, the design is automatically synthesized, using specialized synthesis tools, or manually produced based on simulation and modeling tools. The design is verified to ensure the absence of defects, using specialized verification tools. Verification tools also employ simulation and modeling tools to extract relevant information from the circuit. These stages are common in most IC design processes; however, the individual flow for each circuit is a complex process with many different stages [239]. In this chapter, the individual stages within modern semi-

© Springer Nature Switzerland AG 2022
G. Krylov, E. G. Friedman, *Single Flux Quantum Integrated Circuit Design*,
https://doi.org/10.1007/978-3-030-76885-0_7

Fig. 7.1 Row-based standard
cell placement methodology
with channel routing

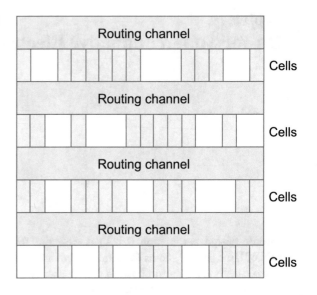

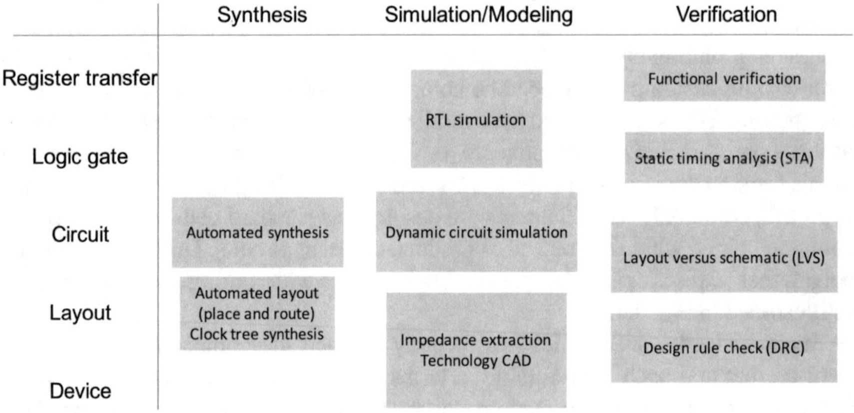

Fig. 7.2 EDA flow for integrated circuits

custom industrial EDA flows are reviewed within the context of superconductive
digital electronics.

To manage complexity, large scale integrated circuits utilize a standard cell-based
design methodology, where well characterized blocks perform specific logic functions. The standard cell design process and related cell libraries for superconductive
integrated circuits are described in Sect. 7.1.

Simulation and Modeling
Simulation and modeling tools are essential for both manual and computer-aided
design of electronic circuits. These tools were among the first tools to be developed.

The design of a complex functional block begins with a description of a circuit at a high level of abstraction. A high level abstraction of a circuit is the register transfer level (RTL) representation, as shown in Fig. 7.2. Tools and techniques for RTL simulation of superconductive circuits are described in Sect. 7.2.

A standard cell consists of circuit elements. To design and characterize each cell, a dynamic circuit simulator is used. In Sect. 7.3, circuit simulators for superconductive circuits are described.

A physical layout of a circuit contains many parasitic impedances. These impedances are not considered during early phases of the design process and can degrade circuit operation. Due to the relative lack of resistive elements in superconductive electronics, the inductance in particular can greatly affect the operation and behavior of superconductive circuits. In Sect. 7.4, techniques and tools for inductance extraction are described.

Synthesis

Automated synthesis tools are a primary capability of a computer-aided design process. These tools enable fast and cost-effective design of large scale circuits, which can be prohibitively complex for manual design processes.

In the automated logic synthesis stage, an RTL description of a circuit is converted into equivalent library elements from a cell library. This process enables fast conversion and optimization of the high level description into circuit components to support the development of large scale systems. Moreover, this flow enables the reuse of existing RTL blocks using different cell libraries and fabrication technologies. Automated logic synthesis for superconductive circuits is discussed in Sect. 7.5.

A synthesized circuit consists of standard cells and connections between these cells. To lay out the actual geometry of the cells and route the connections between the cells according to specific design rules, automated place and route (APAR) tools are required. APAR techniques are reviewed in Sect. 7.6.

Verification

An integrated circuit is a complex system, which can inevitably contain a variety of errors and defects. As the fabrication of an IC is often costly in terms of both resources and time, significant effort is directed at identifying, fixing, and tolerating errors during different stages of the design process.

A multi-gigahertz integrated circuit in particular can exhibit narrow timing tolerances. To ensure correct timing for the logic gates and flip flops, and to prevent race conditions, timing analysis tools are necessary. Timing analysis tools and related techniques are described in Sect. 7.7.

Functional verification is aimed at detecting defects in the RTL description of a circuit. This step utilizes HDL simulators and specific verification methodologies. To detect the errors introduced at the physical layout stage of the design process, design rule checking (DRC) and layout versus schematic (LVS) tools are used. Existing verification techniques and tools for these steps are reviewed in Sect. 7.8. In Sect. 7.9, a summary of this chapter is provided.

7.1 Cell Library Design and Characterization

When the complexity of a circuit exceeds several gates, it is convenient to utilize reusable circuits and blocks (or macrocells). The smallest of these blocks corresponds to the logic gates and flip flops. These blocks—standard cells—are widely used in the design process of both conventional and superconductive circuits.

A manual design process is commonly used in current superconductive electronics. Even with a manual design process, the reuse of standard cells enables the development and characterization of each cell only once, followed by reuse of these cells throughout the circuit design process. In an automated design flow, standard cells are an essential element. In this section, existing standard cell libraries and characterization methodologies for superconductive electronic circuits are described. Many industrial and academic cell libraries have been developed for internal use. Some of these libraries described in the literature are briefly reviewed here.

Early cell libraries were primarily developed for manual or semi-manual design of RSFQ circuits. The RSFQ cell library developed at Stony Brook University is one of the earliest cell libraries developed for RSFQ circuits [241]. The circuits are optimized to provide ease of interconnectivity with other cells in the library. This capability minimizes the redistribution of the bias currents among the connected cells, which can affect the circuit operation and timing characteristics. The timing characteristics of the cells are not provided [241]. Physical layouts are available for most cells, although the layouts utilize a relatively out-of-date 3.5 μm Hypres fabrication process [241]. Most of the cells have been experimentally verified. The Stony Brook cell library remains one of the few open-source RSFQ cell libraries and is a foundation for a significant portion of the circuit development process in modern RSFQ circuits. Another openly available RSFQ cell library was produced at the Ilmenau University of Technology [242] and targets the Fluxonics foundry [243].

The CONNECT cell library was developed in collaboration between Nagoya University and NEC [244]. The cells are also optimized for interconnectivity by minimizing bias current redistribution. The cells are characterized and timing information is extracted without parameterization—the dependence of the cell delays on the neighboring cells and process variations is not described. The physical layouts are targeted for an NEC niobium fabrication process [245].

An RSFQ cell library has been developed at the National Institute of Advanced Industrial Science and Technology (AIST), also in Japan [246]. This library uses a relatively advanced 10 kA/cm^2 AIST fabrication process. Previously developed cells [247] have been scaled and reoptimized to utilize less area while preserving functionality.

An important distinction of modern SCE cell libraries as compared to earlier cell libraries is the focus on automated cell placement and interconnect routing using APAR tools. This focus produces several major requirements. The cell library must satisfy routing rules or allocate sufficient space for routing. The timing characteristics of the cells must be extracted to enable automated timing analysis.

A common approach to the automated placement and routing process in MSI and LSI CMOS circuits is row-based placement. As modern superconductive circuits exhibit limited complexity, the techniques utilized by LSI CMOS circuits are frequently applicable. Modern RSFQ and AQFP circuits adopt a row-based standard cell placement methodology with channel routing, as depicted in Fig. 7.1. The standard cells are arranged in rows, while the space between these rows is reserved for the signal routing channels [240]. While each cell can occupy different area depending upon the function, a uniform cell height is required to produce standard height rows.

An RSFQ cell library for a modern MIT LL SFQ5ee fabrication process [139] has recently been developed [248]. This library supports both RSFQ and ERSFQ bias distribution by separating each cell into multiple components. The core component of the cell contains most of the circuit layout. The bias component contains the relevant bias elements—a bias resistor in the case of conventional RSFQ bias networks and a bias JJ with a large inductor in the case of energy efficient ERSFQ bias networks (see Chap. 13) [49, 50]. Another component of the cell contains tracks for routing PTLs and bias lines, which is intended for use by an automated router.

A methodology for extracting the timing parameters of the standard library cells has been proposed [249]. The cell delays for different combinations of preceding and subsequent cells are extracted in a standard format. These delays can be included within modern industrial CMOS static timing analysis (STA) tools, as further discussed in Sect. 7.7. A methodology is proposed in [160] for parameterizing these standard cells. This methodology allows the cells to be rotated and flipped—common operations during the design process—while correctly maintaining the bias and PTL tracks.

A cell library for the automated layout of AQFP circuits has also been developed [250]. This library utilizes uniform standard cells and is integrated into a semi-custom design flow. Certain steps of the design flow, such as circuit simulation and layout placement, are performed by industrial tools. Other steps, such as retiming, utilize custom tools developed specifically for AQFP.

A bottom-up approach to building an AQFP cell library has been proposed [251]. This library utilizes only four basic elements—branch, buffer, constant, and NOT. All AQFP cells are composed from these elements, greatly simplifying the library design process.

A tool for the automated extraction of the timing parameters of AQFP cells has been developed [252]. This tool extracts setup/hold time characteristics and delay information from parameterized sweeps of different input stimuli by utilizing a dynamic circuit simulation capability. The timing information is included within an industry standard format for use by timing analysis tools.

7.2   RTL Design and Simulation

Electrical circuit simulation exhibits high accuracy but low computation efficiency. When the size of a circuit exceeds several thousand active elements, the circuit simulation process becomes prohibitively long. A solution to this complexity issue is to simulate the circuit at higher levels of abstraction. The register transfer level (RTL) abstraction layer models the transfer of data between registers with functional operations on the data. This abstraction level is often described in a hardware description language (HDL). RTL simulation is technology independent—the logic gates and flip flops described by an HDL can be based on different technologies and circuits. HDL simulation for RSFQ circuits generally differs from the HDL simulation of CMOS circuits due to pulse-based signaling. In RQL and AQFP, where AC clock signals are utilized, a different representation of the signals is necessary. In addition, the extraction of timing parameters suitable for use in HDL models is an important issue. In this section, existing research on HDL for superconductive circuits is reviewed.

Logic simulation of RSFQ circuits was first demonstrated in 1993 [254]. Although this approach enabled behavioral and timing simulation of RSFQ circuits, it utilized a CMOS-oriented model structure specific to a proprietary simulation tool. A PSCAN circuit simulator [255] utilizes an internal HDL language, SFQHDL, to describe the correct circuit behavior. This feature enables automated margin analysis and parameter optimization of RSFQ gates.

RTL-level models of RSFQ circuits using general purpose HDLs were proposed in 1997 [253, 256, 257]. Both Verilog HDL [253] and very high speed integrated circuits (VHSIC) hardware design language (VHDL) [256, 257] can be used. In this approach, gate level HDL models of RSFQ circuits are developed for each logic gate and flip flop. SFQ data and clock pulses are modeled as simple rectangular pulses [258]. An example of HDL simulation of a half adder is shown in Fig. 7.3. The HDL simulation operates with logic values of signals rather than voltages and currents. The internal structure of the gates is not modeled, reducing the computational complexity. Only the internal state of the gates is stored within the models. The

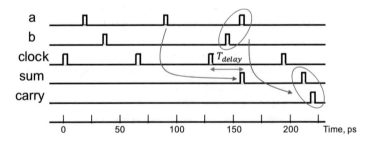

Fig. 7.3 HDL simulation of an RSFQ half adder [253]

behavior of the cell is modeled as a finite state automaton (FSA), where an input pulse causes a transition between states and/or generates an output pulse [177].

RSFQ gates exhibit specific timing requirements. Transitions between the states of an FSA occur within a specific time window. To produce accurate simulation results, timing characteristics are included within the HDL models. These characteristics include the output delay, setup time, and hold time. The timing characteristics are extracted from circuit simulations. Many additional factors exist that can change these characteristics—parasitic impedances, temperature changes, bias current fluctuations, and fabrication process variations. Moreover, these characteristics are frequently data dependent.

Multiple ways to integrate these factors into HDL models have been proposed [177, 259, 260]. The timing characteristics of isolated cells are initially determined. The parasitic impedances are extracted and back annotated into the circuit design flow [177]. The next step is to extract the delay of combinations of gates and more complex circuits [259]. The probabilistic nature of the delay of RSFQ cells is typically modeled as a normal Gaussian distribution [260]. Monte Carlo simulations are performed to determine the delay of the more complex blocks.

Tools and methodologies for the automated extraction of the HDL models and timing characteristics from circuit netlists have recently been proposed [261, 262]. In this process, SFQ pulses are applied to the circuit inputs, allowing different states within the flux storing loops to be identified. This capability enables the automated extraction of an FSA representation of a specific circuit. Critical timing characteristics are also extracted by applying input pulses in a binary search pattern and verifying the output states [262].

A different method for representing signals in an HDL is necessary for RQL and AQFP logic. In RQL, information is represented as the presence or absence of two reciprocal SFQ pulses, while the clock is a multiphase AC sinusoidal signal. The clock signal in an RQL VHDL model is composed of three regions based on the magnitude and direction—positive pulse propagation, negative pulse propagation, and no propagation [108]. The total change in phase of the JJs within the RQL gates is zero. The change in phase is only nonzero during the time between the arrival of the positive and negative SFQ pulses comprising the signal. This property enables a natural translation of RQL signals into an HDL. Logic "zero" corresponds to the absence of activity. In the case of a logic "one," positive and negative SFQ pulses correspond, respectively, to the positive and negative edges of a conventional CMOS-like signal [108].

A similar approach to RQL is used in HDL models of AQFP logic. The direction of the excitation current determines the HDL state. A negative current corresponds to logic "zero," a positive current to logic "one," and the absence of a current corresponds to the high impedance Z state [263].

The SystemVerilog language has recently been proposed as an HDL model of RSFQ and AQFP circuits [263, 264]. SystemVerilog models provide multiple benefits, among which are compatibility with industrial tools, modular reusable design, and integration with verification methodologies. Moreover, this approach enables HDL simulation of hybrid RSFQ/AQFP systems.

7.3   Circuit Simulation

In this section, circuit simulators capable of analyzing circuits with Josephson junctions are reviewed. Dynamic circuit simulators for superconductive electronics operate with voltages, currents, and phases at different nodes and can produce highly accurate results, while exhibiting high computational complexity. An example waveform of a circuit simulation of a JTL is shown in Fig. 7.4, where the input and output voltages of a JTL with two JJs (stages), connected to a clock signal, are shown.

Two different methods are used to simulate superconductive circuits. In one approach, the node voltage is the fundamental variable of a dynamic simulator. This approach is commonly adopted in dynamic simulators for conventional electronic circuits based on the original SPICE simulator [265]. Among these SPICE-based tools for SCE are JSPICE3 [266], JSIM [267], and WRspice [268]. These simulators are commonly used and include a variety of techniques to increase simulation speeds for circuits based on Josephson junctions [268].

Certain SPICE-like simulators support circuits with JJs by using a physics-based device model. Verilog-A is an HDL (see Sect. 7.2) commonly used to describe an electronic device based on the physical equations characterizing the behavior of the device. A Verilog-A model of a Josephson junction is typically based on the RCSJ model, described in Chap. 2. The Verilog-A model is relatively simple, containing only a few expressions. The device model enables simulation of JJs with the same industrial simulators used for CMOS circuits, such as HSPICE and Spectre. The use of an external model within a regular circuit simulator, however, can be computationally slower than a model embedded within the tool due to the absence of enhancements specific to JJ-based circuits [269].

Among the advantages of this type of simulator is compatibility with existing circuit simulation tools. Another important benefit of this capability is simulating

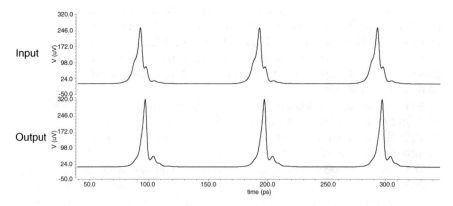

Fig. 7.4 Circuit simulation of a two-stage JTL. The voltages across the input and output JJs are shown

both superconductive and semiconductor devices together (mixed technologies). This capability is beneficial in SCE, for example, for RSFQ/CMOS interface circuits. Among the disadvantages of existing simulation tools is the difficulty of performing DC analyses, as the DC operating point depends upon the phase of the JJs. Modern versions of WRspice [268] and JoSIM [270] support DC analysis based on the phase.

A different approach to simulate superconductive circuits is to treat the phase of the node rather than the voltage as a fundamental variable. The phase is a natural parameter characterizing the state of a JJ, allowing the circuit state, signals, and behavior to also be characterized by the phase. The most widely used simulators utilizing this approach are PSCAN [255, 271] and PSCAN2 [272]. These tools only consider superconductive circuits and support additional forms of analyses. Another recently developed simulator, JoSIM [270], operates with both voltages and phases. No studies, however, exist comparing these approaches in terms of relative computational speed and/or accuracy [269].

An important feature of a superconductive circuit simulator is whether the microscopic Werthamer model of a Josephson junction [273, 274] is supported. Although the RCSJ model is sufficiently accurate for most modern digital superconductive circuits, this model targets digital circuits with externally shunted JJs, where the shunt resistance and inductance are small. In addition, the variation of voltage across a JJ, on a picosecond time scale, is typically assumed to be small in comparison to the gap voltage [275]. Significant differences exist between the microscopic Werthamer model and the RCSJ model for junctions with high damping [75]. The microscopic Werthamer model enables simulation of deeply scaled unshunted Josephson junctions with a high critical current density [274]. Currently, PSCAN2 [272] and JoSIM [270] can use embedded microscopic JJ models.

7.4 Inductance Extraction

On-chip inductance is a critical design parameter for superconductive circuits. Small variations in the inductance can produce incorrect circuit operation. While for simple structures the inductance can be roughly estimated based on fabrication process information, most practical circuits include inductors of complex geometry. To design circuits with these inductors, accurate extraction tools are necessary. In this section, these tools and relevant techniques are described.

Inductance extraction tools can be broadly separated into two categories—2-D and 3-D [269]. 2-D extraction tools are typically significantly faster and significantly less accurate than 3-D tools. As described in an important review on inductance extraction [269], 2-D methods are not commonly used in the design of modern superconductive circuits.

An intermediate step between 2-D and 3-D simulation is 2.5-D [276] or planar 3-D [269] simulation. This type of 2.5-D analysis is used in the Sonnet field solver [277, 278]. In this 2.5-D simulation system, a three-dimensional circuit geometry is separated into conducting surfaces, which are partitioned into two-dimensional cells. Field equations are solved for these cells based on a surface impedance model using the method of moments [279]. This type of analysis is faster than 3-D analysis [269] while providing good extraction accuracy for most structures. This simulator, however, exhibits significantly lower accuracy for narrow submicrometer lines, expected in next generation superconductive circuits [276].

The Ansys HFSS tool is also used to extract impedances in superconductive circuits [280]. This tool utilizes three-dimensional cells and the finite element method [281]. A distinctive advantage of complex microwave simulators such as HFSS and Sonnet is the capability to extract frequency dependent effects.

A commonly used 3-D inductance extraction tool in superconductive electronics is FastHenry [282]. This field solver was originally developed and widely used for conventional CMOS circuits [283]. In FastHenry, a conductor is divided into segments and subdivided into filaments—the partial element equivalent circuit method (PEEC). Filaments and terminal sources form an equivalent circuit from which a complex impedance matrix is extracted. This tool, with additional modifications, is capable of simulating superconductive structures. In [284], an additional term corresponding to the kinetic inductance is added to the basic governing equations. In [285], the London equations and two-fluid model are used to enable support for superconductive structures. FastHenry produces highly accurate results for simple structures; the tool, however, is computationally expensive. Extracting the inductance of complex geometries requires a fine mesh composed of many filaments and can require a prohibitively long computational time.

One of the most popular inductance extraction tools suitable for larger and complex structures is InductEx [286]. Initial versions of this tool are based on a FastHenry engine. InductEx utilizes a novel segmentation algorithm, where the edges of the geometry are divided into finer segments than the simpler regions. Multiple modifications have also been introduced to improve the speed of the analysis process. A novel field solver has been developed utilizing cuboid segments [287]. The latest versions utilize a tetrahedral mesh and other enhancements to improve both accuracy and speed [288].

A different methodology is used in the 3D-MLSI tool [289, 290]. In this tool, individual currents are derived from stream functions. London and Maxwell equations are described in terms of these stream functions, and the resulting expression is solved using the finite element method [291]. This approach is computationally efficient and enables inductance extraction of complex structures [292].

7.5   Logic Synthesis

During logic synthesis, the behavioral description of a circuit is converted into a gate level netlist for a specific cell library. Multiple differences exist in RSFQ circuits as compared to conventional CMOS circuits, requiring modifications to existing tools [153, 154]. In this section, existing approaches to logic synthesis for superconductive electronics are discussed.

7.5.1 Logic Representation

One of the first methodologies for the automated synthesis of RSFQ circuits is the top-down binary decision diagram (BDD) methodology [293]. A BDD is an acyclic directed graph consisting of decision nodes and terminal nodes [294]. A boolean function can be represented as a binary decision tree, as shown in Fig. 7.5a, where each variable corresponds to a decision node, while the terminal nodes correspond to the value of the boolean function. A different path in the graph exists for each combination of variable values X_1, X_2, and X_3. A BDD, shown in Fig. 7.5b, is a reduced representation of a binary decision tree where the redundant nodes and edges are omitted.

In the BDD-based synthesis methodology, the cell library contains only one logic gate—a binary switch [293]. In RSFQ technology, this switch is based on a D flip flop which does not require significant area. Other elements in the cell library provide connections between the switches. BDD transformations, however, exhibit high computational complexity and are therefore not feasible for large scale circuits.

A commonly used academically developed synthesis tool, ABC [295], is applicable to RSFQ circuits. ABC converts the behavioral description of a circuit into an intermediate representation—the and-inverter graph (AIG). AIG is an acyclic directed graph consisting of conjunction nodes, terminal nodes, and edges that can

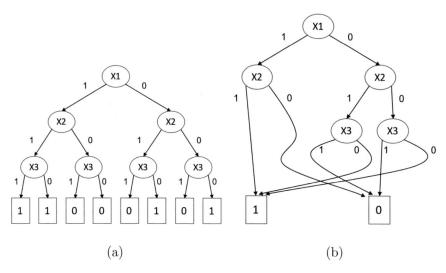

(a)   (b)

Fig. 7.5 Graph representations of a circuit used in the synthesis process: (**a**) binary decision tree and (**b**) binary decision diagram (BDD)

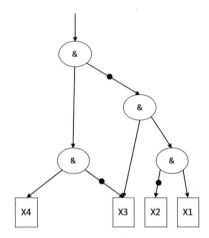

Fig. 7.6 And-inverter graph (AIG). The black dots on the edges denote the inversion operation

contain the inversion operation. An example AIG is shown in Fig. 7.6, where each node represents the conjunction operation on the corresponding child nodes, and the inversion operations are represented by black dots. Specific transformations are performed on the AIG representation of the circuit [296] to enhance the synthesized circuit. The ABC tool is widely used as a platform for synthesis optimization.

A modification of the AIG representation of a boolean function is the majority-inverter graph (MIG) [297]. In MIG, three input majority nodes are used rather than conjunction nodes. This format enables a more natural representation of logic circuits with efficient majority elements, such as QFP [298] and DSFQ [48] circuits. The MIG representation also supports specific optimizations that exploit the properties of the majority function [299].

AQFP logic is also compatible with standard CMOS synthesis tools, such as yosys [300]. QFP logic exhibits an efficient inversion operation and limited fanout. Additional transformations of the netlist are therefore necessary [301]. A synthesis methodology for AQFP circuits has been proposed [302]. In this methodology, a regular And-Or-Inverter (AOI) graph is converted into a majority-minority graph, similar to the MIG. As AQFP technology features efficient majority gates, this approach reduces the number of JJs as compared to an AOI representation, as well as lowers delay due to a decreased logic depth.

7.5.2 Path Balancing

Multiple changes are necessary for ABC to support the synthesis of RSFQ circuits, first proposed in [303]. As previously described, most logic gates in RSFQ technology are individually clocked [52]. If different inputs of a gate exhibit a different logic depth, an erroneous output is produced. To balance the depth of all

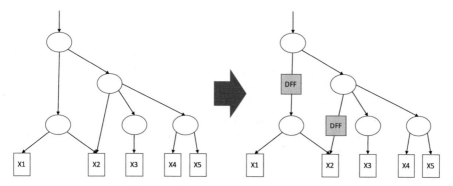

Fig. 7.7 Insertion of path balancing D flip flops into RSFQ logic paths

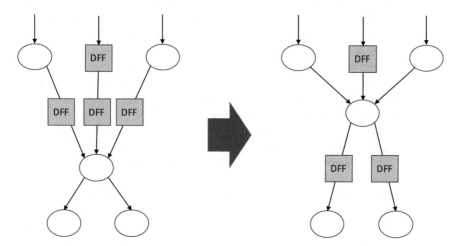

Fig. 7.8 Retiming of a logic path with path balancing DFFs

of the gate inputs, path balancing D flip flops are inserted. As shown in Fig. 7.7, for every gate in a circuit, DFFs are inserted into an input path that exhibits a smaller logic depth. These flip flops operate as dummy gates, performing no functional operation other than delaying a data pulse by a single clock period. This process is referred to as path balancing.

In a complex RSFQ logic path, many path balancing DFFs are needed, requiring significant additional area. Multiple techniques and algorithms have been developed to reduce the number of these DFFs [304]. One such technique is retiming [305, 306] where the logic gates and path balancing DFFs are rearranged to reduce the number of flip flops. The effect of retiming on the total number of required path balancing DFFs is shown in Fig. 7.8. A technique to convert a CMOS circuit netlist into an RSFQ netlist has also been demonstrated [307]. This technique utilizes retiming to reduce the number of flip flops.

7.6   Layout Synthesis

The output of the synthesis process is a symbolic description of a synthesized circuit [51]. At this stage, the circuit is composed of library cells and connections between these cells. The physical topology of the cells is laid out—the geometric descriptions of elements are assigned to specific locations on an IC. The necessary connections between cells are routed—interconnect elements are placed to establish these connections. In this section, automated placement and routing (APAR) tools and related algorithms for superconductive circuits are described.

APAR for RSFQ

One of the first automated APAR methodologies for RSFQ system was developed [172] to lay out an 8-bit general purpose RISC processor composed of 20,000 cells. This methodology only utilizes PTL interconnect for routing and an H-tree topology for the clock distribution network. The placer is based on the Fiduccia-Mattheyses (FM) heuristic [308] to recursively partition a circuit while minimizing the number of connections between the resulting partitions. By placing the connected cells near each other, the maximum wire length and therefore delay are reduced, while increasing the maximum clock frequency. The cells are synchronized by an H-tree clock distribution network connected to each cell. Routing in this methodology can be performed by any industrial CMOS routing tool [309].

A routing tool specifically targeted for RSFQ circuits is proposed in [310] and used to route an 8-bit microprocessor. Only PTL interconnect is utilized for the routing. This tool is based on the A∗ algorithm [311], widely used both for CMOS routing and, in general, for minimizing the cost of a path within a graph. The tool also adjusts the cost of these paths to decrease the number of vias and corners along a path.

The delay of the JTL interconnect strongly depends upon variations in the manufacturing process characteristics. Variations in the bias current, JJ size, or inductance can change the delay by several picoseconds. This effect can be unacceptable in a multi-gigahertz system. PTL interconnect also exhibits small variations in the delay which primarily depend upon the length of the line [107, 146]. It can therefore be desirable to insert certain delays within a circuit by increasing the length of the PTL rather than inserting a JTL delay element. In [312], this property is exploited in a routing tool based on integer linear programming (ILP) [313]. This technique is similar to wire snaking, commonly used in CMOS circuits for delay balancing [234].

An extension of this work is described in [314]. Simulated annealing (SA) [315] is used to decrease the routing time as compared to the ILP approach. Segments of the PTL interconnect—delay matching elements—are inserted to balance the delay of the different paths. The SA algorithm is also used for automated placement, where the length of the wires and the delay matching elements are minimized. In [316], this approach is improved by rearranging the delay matching elements. This technique reduces the minimum width of the routing channels.

An APAR methodology and tool are proposed in [178, 179] based on an HL-tree clock distribution network. In this approach, the cells are grouped by the increasing logic level. Cells within each group are abutted. Both the area and the total wire length are decreased. In this methodology, the open-source qrouter tool [317], a tool based on the Lee maze routing algorithm [318], is used for routing, while the SimPL algorithm [319] is used for global placement.

The delay and area of a PTL interconnect is greater than a JTL interconnect for short distance routing. This property is due to the relatively large delay of the driver and receiver required to interface with the passive striplines. A mixed approach for routing RSFQ circuits has also been proposed [320], utilizing both JTLs and PTLs.

A place and route methodology for DDST RSFQ circuits has been proposed [321]. This methodology utilizes industrial CMOS-based tools to synthesize and lay out asynchronous RSFQ circuits.

APAR for AQFP

AQFP circuits are typically placed in a row-based standard cell topology. Each row corresponds to a different logic level and is synchronized by a different clock phase.

AQFP circuits utilize a different type of interconnect from RSFQ logic. The current waveforms used for signaling in this technology propagate within the regular metal wires, similar to CMOS. Multiple restrictions exist when routing these wires, as long interconnect segments introduce attenuation. In [322, 323], an APAR methodology for AQFP circuits is proposed. In this approach, a genetic algorithm (GA) [324] is used to place the cells while reducing the number of long interconnect segments. Buffers are inserted into the remaining long lines. The left-edge algorithm [325] is used for channel routing. This methodology has been used to automatically place and route a 16-bit AQFP adder [322].

7.7 Timing Analysis

Superconductive electronics operate at extremely high clock frequencies and exhibit small gate delays [151]. The timing analysis process in these circuits is essential for high speed operation. In this section, timing analysis and related techniques for superconductive electronics are reviewed.

Conventional CMOS circuits utilize different techniques for timing analysis. Dynamic timing analysis (DTA), which operates at an abstraction layer above dynamic circuit simulation, simulates a system at the behavioral level. Although this approach can produce accurate results, it is computationally expensive. This analysis is similar to verification and is therefore described in greater detail in Sect. 7.8.

Static timing analysis (STA) is much faster than DTA. During the static timing analysis process, the expected delay of different gates and logic paths is compared to the minimum and maximum allowed delays. The delay of each standard cell—gates and flip flops—is extracted and compiled into a lookup table (LUT). Certain cell parameters are evaluated to consider manufacturing variations in the fabrication process. The delay of the interconnect lines as well as the dependence of the delay on the load and fanout is also considered within the LUTs. These tables allow quick estimation and detection of timing violations without requiring the circuit to be simulated on a nodal basis.

7.7.1   Timing Constraints

Multiple differences exist between RSFQ logic and conventional CMOS circuits that prevent the straightforward adoption of existing STA CMOS tools to RSFQ. One of these differences is signaling for clock and data. RSFQ circuits utilize SFQ voltage pulses for signaling. Although these pulses exhibit a quantized area Φ_0, the magnitude and duration of these pulses can vary for different gates. It is therefore difficult to determine the precise moment when a pulse arrives or is generated—an important issue in timing analysis. In certain conditions, the peak amplitude of an SFQ pulse does not correspond to the moment of switching. Moreover, transient noise in RSFQ circuits can be similar to an SFQ pulse, with the primary difference being the area of the waveform.

One method to determine the timing of a pulse is to monitor the phase of the input/output JJs within a gate. An input pulse incoming to a gate produces a 2π change in the phase of the input JJ. Conversely, the output pulse produces the same change in the phase of the output JJ. The duration of the complete 2π change varies depending upon the damping and bias conditions. In general, however, it is characterized by a steep slope and a specific settling time. A commonly used technique to set the precise moment of switching is to use a specific fraction of the 2π change, e.g., 75% [269].

This technique is shown in Fig. 7.9, where both the voltage and phase of a switching JJ are shown. These fractional changes in the phase of a JJ are less dependent on the circuit parameters and settling time, exhibiting smaller variations, and are therefore appropriate for timing analysis. This approach, however, cannot be used if the internal structure of the gate is not accessible. In this case, the phases of the interconnect JJs can be used. In the case of JTL interconnect, for example, the phase of the first and last junction of the interconnect can serve as a temporal reference. For PTL interconnect, the phase of the driver and receiver JJs can be used.

Differences in the gate timing characteristics extracted from the phase change of the output junction of a gate rather than the peak output voltage can reach several picoseconds. While both methods are used, it is important to extract timing characteristics in a consistent manner.

Another important distinction of RSFQ logic is that most logic gates require a clock signal. Due to this feature, the logic cells are treated as sequential elements. Standard CMOS timing concepts, such as the setup time or hold time, are therefore redefined for RSFQ circuits [253, 326].

In general, multiple critical timing constraints exist in RSFQ logic circuits [262]. These timing constraints are illustrated in Fig. 7.10. One constraint is the minimum

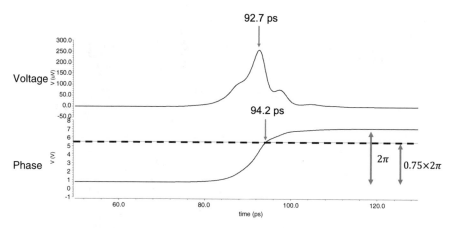

Fig. 7.9 Pulse timing based on a phase change of a JJ. The voltage and phase across a JJ are shown during switching. The dashed horizontal line denotes a 75% increase in the phase of the JJ

separation time between the input pulses and the clock pulse. If an input pulse arrives too late, the clock signal produces an output based on the incorrect state of the gate, as shown in Fig. 7.10a. This clock-after-data [249] timing constraint is similar to the CMOS setup time constraint [101]. Another critical timing constraint is the minimum separation time between the clock pulse and any incoming input pulses during the next clock period, as shown in Fig. 7.10b. If an input pulse arrives too early, the state of the logic gate does not change, producing an error. This data-after-clock [249] separation time is similar to the CMOS hold time constraint [101].

A different set of critical timing constraints exists for asynchronous circuits, which is related to the minimum separation time between input pulses. An example of a circuit function affected by this timing constraint is the merger. If the separation time is violated, one pulse is produced rather than two, as shown in Fig. 7.10c. These critical timing constraints are different for each gate, and some of these constraints do not exist for certain gates.

RSFQ circuits can be abutted to each other, with the output of a gate directly connected to the input of another gate. In this case, the cell delay depends upon both the preceding and subsequent cells. A major difference between CMOS and RSFQ circuits for timing analysis is the two different types of interconnect—PTL and JTL—used in RSFQ. The interconnect in CMOS is composed of metal lines directly connecting the logic gates [101, 327, 328]. JTLs in RSFQ circuits are active elements and require abutment to the cells driving and receiving SFQ signals. JTLs can therefore be characterized in a similar manner to any other logic cell within a cell library. PTLs are composed of striplines and require a driver and receiver to operate. In this case, the driver and receiver can be characterized as active logic gates. These gates are connected by a (typically) long passive metal stripline or a microstripline. The interconnect delay for a PTL is related to the speed of light (in the medium) in these long metal lines [23].

Fig. 7.10 Critical timing
constraints in RSFQ circuits,
(**a**) setup time, (**b**) hold time,
and (**c**) separation time

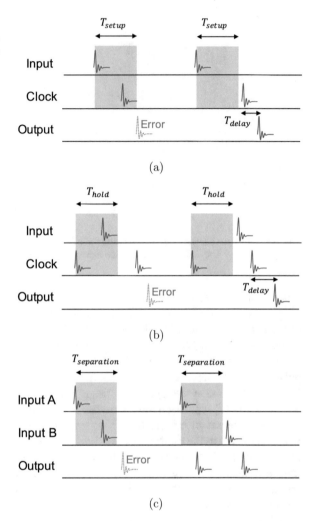

7.7.2   STA Techniques

Specific algorithms and techniques exist for static timing analysis of RSFQ circuits.
A statistical approach for timing analysis has been proposed [329]. In this methodology, variations in the output delay are based on a statistical distribution of different
fabrication parameters. In [330], algorithms for estimating the path delay and timing
slack of different logic paths are presented. Based on this timing slack, the minimum
system clock period is determined. An STA tool is described in [331]. In this tool,
the critical paths are detected and the timing slack is determined based on the worst
case cell delays and the length of the PTL interconnect between cells.

A similar definition of a critical timing window exists for AQFP circuits [263]. Both data and clock are represented as current waveforms. This feature enables greater similarity with CMOS signaling and related timing constraints, increasing compatibility with standard CMOS timing analysis methodologies [101].

7.8 Verification and Testability

The description of a circuit is modified at every step of the design flow (see Fig. 7.2). From logic synthesis to layout and fabrication, many different tools and steps introduce changes into a circuit and can therefore affect functionality or introduce defects. To negate the risk of errors, verification is necessary. In this section, verification of both the logical circuit structure and the physical layout is discussed. Built-in self-test (BIST) techniques, which assist in verifying the operation of a fabricated circuit, are also reviewed.

The operation of a circuit is initially described in a target specification. These specifications are manually converted into an RTL description. To verify the correctness of this conversion process, confirming that the RTL matches specifications, functional verification is performed. As the RTL design process for superconductive circuits is similar to the design process of conventional CMOS circuits, many CMOS compatible techniques can be reused. Until recently, the complexity of superconductive circuits was not sufficiently high to require rigorous functional verification. For the 8-bit FLUX-1 processor, composed of approximately 66,000 JJs, functional verification was partially performed for individual blocks using VHDL [332]. These blocks were also individually fabricated and tested before integration into a single circuit. HDL verification has also been used for AQFP [263]. For prospective large scale superconductive processors, more rigorous functional verification is necessary.

The synthesis process converts an RTL description of a circuit into an optimized netlist suitable for layout. During this process, defects can be introduced by the synthesis tools. To detect these errors, formal equivalence checking (FEC) is performed [333]. FEC techniques are used to formally verify in all practical cases that the RTL description is equivalent to the netlist. Alternatively, two netlists can be compared before and after certain transformations. A logic equivalence checking methodology and tool has been developed for RSFQ circuits [334]. This tool verifies any fanout restrictions and the correctness of the path balancing process. A combined formal and functional verification tool has also been proposed for RSFQ circuits [335]. This tool uses formal methods to verify fanout constraints and the correctness of the path balancing process. Upon completion, the circuit is functionally verified using the industry standard Universal Verification Methodology (UVM) framework. The UVM framework utilizes SystemVerilog features to enable a quick design process and to facilitate reuse of the verification environment [336].

Once the synthesized netlist is laid out, additional defects can be introduced. Specialized verification steps are therefore required. Design rule checking (DRC) verifies that a laid-out circuit does not violate the geometric spacing rules of the fabrication process. This verification process evaluates the distances, sizes, and other geometric properties of the different layout structures. A layout versus schematic (LVS) check confirms the equivalence of the laid-out circuit with the initial logic netlist. This process extracts a netlist from the physical layout and compares this netlist with the initial logic netlist. Both DRC and LVS are used for superconductive circuits [337, 338]. For these verification steps, many compatible CMOS methods and tools are reused.

Defects can be introduced into a verified circuit during the fabrication process. Parameter variations can affect the timing characteristics and lead to incorrect operation, particularly in multi-gigahertz superconductive VLSI systems. To evaluate the fabricated circuits, design for testability (DFT) features are introduced into superconductive circuits [53, 54]. These features enable easier detection and localization of errors during the testing process. DFT methodologies for RSFQ circuits are reviewed in greater detail in Chap. 16.

7.9 Conclusions

An EDA flow for superconductive circuit design and analysis is described in this chapter, including a general background and issues specific to superconductive circuits. For each step of the design flow, tools and algorithms are discussed, and sources for more detailed information are provided. Existing standard cell libraries for superconductive circuits and related cell library design and characterization techniques are reviewed. For the automated synthesis process, methodologies and algorithms are described for logic synthesis and automated place and route. For the simulation and modeling process, RTL simulation using HDLs and different dynamic and static circuit simulators, as well as inductance extraction tools, is described. For the verification process, timing analysis methodologies and related timing constraints suitable for modern superconductive circuits are discussed, and verification approaches are reviewed.

Existing EDA tools and techniques for superconductive electronics are immature as compared to CMOS EDA tools. Significant research efforts are, however, directed at improving and developing novel algorithms and design methodologies that target superconductive circuits. The effectiveness of these tools is improving to enable large scale superconductive systems.

Chapter 8
Compact Model of Superconductor-Ferromagnetic Transistor

The superconductor-ferromagnetic transistor (SFT) is a novel cryogenic device with the potential to greatly enhance traditional single flux quantum circuits. Since SFT devices are under active development, compact models are necessary to include this device in the simulation of novel circuits. In this chapter, a simplified compact model of a three-terminal SFT device is proposed. The model fits the general I-V characteristics of existing devices with 7.4% mean absolute error, while also capturing the transient behavior of the device. The model has been implemented in Verilog-A and simulated in Cadence Spectre. The proposed model enables the simulation of SFQ circuits containing SFT devices, and is reconfigurable to support developments in SFT technology [45].

8.1 Introduction

Despite recent developments in single flux quantum technology, one major drawback is the lack of a fast and dense memory [47]. An unusual characteristic of SFQ technology is the absence of a three-terminal device providing good input-output isolation and controllable switching behavior. SFQ circuits are composed of two-terminal Josephson junctions. The introduction of a three-terminal device would greatly enhance circuit flexibility and support the development of novel circuits, particularly memory systems.

One novel family of three-terminal devices is superconductor-ferromagnetic transistors, previously described in Sect. 2.5.4.1. These devices are composed of stacks of superconductive (S), ferromagnetic (F), and insulator (I) layers where the electrical properties depend upon the arrangement of the layers within a stack and the number of terminals. In a three-terminal $SFIFSIS$ device, the critical current between the SIS acceptor terminals is modulated by the current supplied to the $SFIFS$ injector terminal.

© Springer Nature Switzerland AG 2022
G. Krylov, E. G. Friedman, *Single Flux Quantum Integrated Circuit Design*,
https://doi.org/10.1007/978-3-030-76885-0_8

To enable circuits using this novel SFT, a closed-form model of this device is necessary [269]. Previously published theory characterizing the operation of these devices [88] utilizes expressions that are overly complex for circuit simulation. A closed-form, computationally efficient model of a three-terminal $SFIFSIS$ device suitable for circuit simulation is described here.

In Sect. 8.2, the compact model of the $SFIFSIS$ device is described. In Sect. 8.3, the model is verified against experimental data. In Sect. 8.4, some conclusions are offered.

8.2   Compact Model of the SFT Device

The compact model of an SFT device is described in this section. In Sect. 8.2.1, the operation of a three-terminal SFT device is presented, and an equivalent electrical circuit is proposed. In Sect. 8.2.2, expressions describing the critical current and superconducting energy gap are discussed, and a simplified closed-form expression is described. In Sect. 8.2.3, approximations of the gain and threshold voltage are presented. In Sect. 8.2.4, the reactive parameters of the device are discussed. In Sect. 8.2.4, the asymmetric characteristics of the device are discussed.

8.2.1   SFT Device Operation

Physical principles and arrangement of layers of the $SFIFSIS$ device are described in Sect. 2.5.4.1. The SIS acceptor with a shared S_2 layer generally exhibits properties similar to regular SIS devices, such as a Josephson junction. A Josephson junction is commonly characterized by a resistively and capacitively shunted junction (RCSJ) model, as discussed in Sect. 2.4.4. The equivalent circuit for this model is shown in Fig. 8.1a. In this model, a junction is represented by an ideal Josephson element connected in parallel with a resistor and capacitor.

In a model of an SFT device, the existing RCSJ JJ model should consider the novel behavior caused by the injector stack. The equivalent circuit of a three-terminal SFT device is shown in Fig. 8.1b. The available experimental data describing the $SFIFS$ injector, such as the linear I-V characteristics of the stack, suggest that the injector current I_i does not exhibit Josephson behavior due to the presence of an exchange field in the ferromagnetic layers [88, 89]. The $SFIFS$ injector exhibits a linear current-voltage characteristic and is therefore represented as a resistor.

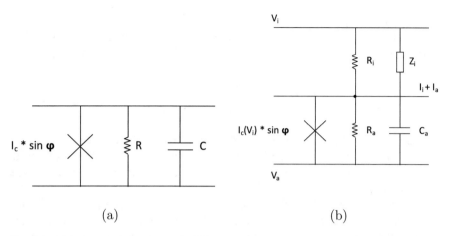

Fig. 8.1 Equivalent electrical circuit, (a) RCSJ model of Josephson junction, and (b) model of a *SFIFSIS* device

8.2.2 Critical Current and Suppression of Superconductive Energy Gap

An expression for the dependence of the critical current on the injector voltage is described in [88]. The integral equations constituting this solution are solved numerically, and are therefore not incorporated into a closed-form, computationally efficient expression suitable for circuit simulation.

Critical current suppression in an SFT is caused by suppression of the superconductive energy gap, which depends upon the injector current, and, consequently, the injector voltage. The energy gap is abruptly suppressed when the voltage reaches a threshold voltage V_{th}.

This dependence resembles a bell function (8.1),

$$f = \frac{1}{1 + |\frac{x-c}{b}|^{2a}},$$ (8.1)

where b is the threshold voltage. The slope of the $\Delta(I_i)$ dependence determines the gain G of the device. Therefore, $2a$ in (8.1), which determines the slope of the curve, is equal to G.

While the critical current of the device is weakly dependent on secondary parameters, the general shape of the $I_c(I_i)$ curve is due to the shape of the $\Delta(I_i)$ relationship. A simplified closed-form expression of the $I_c(I_i)$ relationship is

$$I_c = \frac{\kappa}{1 + |\frac{V_i}{V_{th}}|^{G}},$$ (8.2)

where κ is a fitting parameter.

The acceptor junction model is based on the RCSJ model, where the quasiparticle current-voltage relationship is approximated as a piecewise linear function,

$$I_N(G) = V \times \begin{cases} G_{SG}, & |V| < V_g \\ G_N, & |V| > V_g, \end{cases} \tag{8.3}$$

where G_{SG} and G_N are, respectively, the subgap and normal conductance. In the RCSJ JJ model, a basis for the proposed SFT model, the ratio of the subgap to normal conductance is ~ 0.05, and the conductance gradually transitions from G_{SG} to G_N around V_g.

The injection-dependent dynamic change of the gap voltage associated with the gap suppression is included in the model. The spread of the transition from the normal region to the subgap region is also scaled by the same factor. The ratio of the subgap to normal conductance is assumed to be approximately 0.05, as in the standard Josephson junction model, due to the current lack of experimental data characterizing an SFT.

8.2.3 Gain and Threshold Voltage Model

One primary parameter characterizing a three-terminal SFT device is the ratio of the injector resistance $R_{T(i)}$ and acceptor resistance $R_{T(a)}$. This $\frac{R_{T(i)}}{R_{T(a)}}$ ratio affects both the threshold voltage V_{th} and the gain G of an SFT device.

Both $V_{th}(\frac{R_{T(i)}}{R_{T(a)}})$ and $G(\frac{R_{T(i)}}{R_{T(a)}})$ are numerically characterized in [88]. To provide a closed-form model, however, the dependence is assumed to be approximately linear between the ratio of 1 and 15, gradually changing from 1 to 4.5 mV. The gain of the device also varies linearly, increasing for larger resistance ratios.

Both of these dependences are incorporated within the model. However, as SFT fabrication processes are currently immature, the model can be adjusted manually, allowing the estimated gain and threshold voltage to be fitted to experimental data.

8.2.4 Reactive Parameters of the Injector

As the injector of the device is a complex stack of metal layers, ferromagnetic layers, and an insulating layer, the behavior of the injector is not completely resistive. With the DC-biased injector stack, the acceptor behaves as a current-controlled Josephson junction. Capacitive and inductive effects within the injector affect the electrical properties of the device, particularly during transient switching. The injector impedance Z_i is included in the equivalent circuit shown in Fig. 8.1b. The capacitance and inductance of the injector depends upon the device geometry and can be estimated or measured experimentally.

The capacitance of the injector multilayer stack is included within the model to improve the accuracy of the dynamic transient analysis process. The capacitance is modeled as a parallel plate capacitance between the two ferromagnetic layers within the injector. The capacitance depends upon the area of the injector, relative permittivity of the insulator material, and thickness of the acceptor insulator layer. The relative permittivity is assumed to be approximately 9.1 for aluminum oxide, and the insulator thickness is assumed to be approximately 2 nm. For a typical SFT device, these parameters produce an injector capacitance of approximately 2 pF. The model supports modification of the parameters based on the device geometry. The introduction of an injector capacitance negligibly affects the acceptor properties. The injector capacitance can however be important for the circuit driving the injector.

8.2.5 Asymmetry Parameter

The support for asymmetry of the SFT $I_c - I_i$ relationship with respect to the injection current is included within the model. In certain SFT devices, this dependence is asymmetric with respect to zero injection. Asymmetry is attributed in [339] to current compensation at the edge of the middle superconductor layer. Opposing acceptor and injector currents lead to an increase in the critical current of the acceptor. This effect can be reduced by decreasing the lateral dimensions of the acceptor. To model larger devices, however, this asymmetry is introduced into the model by modifying the expression for the dependence of the critical current on the injector voltage. The asymmetry offset in the model is manually adjusted for each device based on experimental data. As compared to the variations in the $I_c - I_i$ characteristics, the variations in the asymmetry offset are negligible. The model therefore supports the simulation of asymmetric devices during the circuit design process.

8.3 Model Verification

The proposed model has been integrated into an existing RCSJ Verilog-A model of a Josephson junction [266], evaluated within the Cadence Spectre simulator, and compared with experimental data reported in [88]. The I-V characteristics resulting from a DC analysis are shown in Fig. 8.2.

From the DC analysis, the model accurately captures the behavior of the device. The input parameters of the proposed model are the device geometry, gain, threshold voltage, and critical current without injection. The gain and threshold voltage are estimated by the model or manually adjusted based on available experimentally

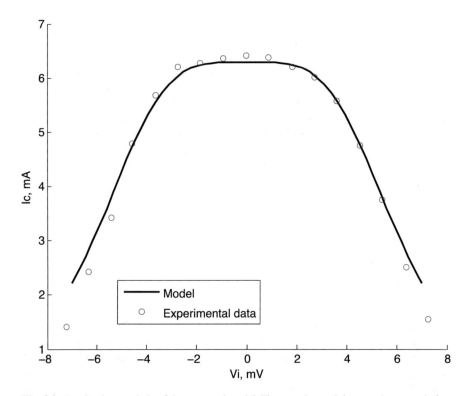

Fig. 8.2 $I_c - V_i$ characteristic of the proposed model. The experimental data are shown as circles

measured parameters. The simulation accurately describes the I-V characteristics of existing devices [85, 88, 340] with a mean absolute error of 7.4%.

To date, no experimental transient measurements of SFT devices exist in the literature. A model characterizing the transient behavior has therefore not been compared to experimental data. This model, however, can be compared to the expected behavior of a Josephson junction with the critical current modulated by the injection current. A simulation of this behavior is depicted in Fig. 8.3. In this transient simulation, the acceptor junction of the model is connected to the bias current source and biased at $I_b = 0.7 \cdot I_c$ [50]. This current is insufficient to switch the acceptor junction into a resistive state. The injector current I_i, initially zero, gradually increases to 1.4 mA, corresponding to an injector voltage of 5 mV. After the injector voltage is applied, the acceptor junction continuously switches, generating a series of SFQ pulses. This switching behavior is due to suppression of the acceptor critical current to approximately 66% of the original critical current. After the injector voltage is reduced, the acceptor critical current is restored to the current level without injection, terminating the switching process (see Fig. 8.3d). A simulation of this behavior accurately captures the expected transient behavior

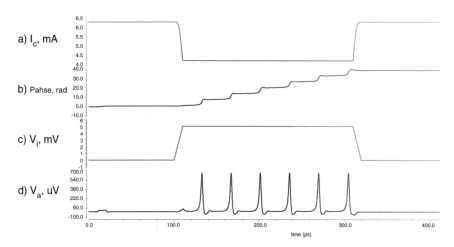

Fig. 8.3 Transient response of an SFT device based on the proposed model, (**a**) critical current of the acceptor, (**b**) phase of the acceptor, (**c**) injector voltage, and (**d**) acceptor voltage

of the device and confirms the application of the proposed SFT model to the SFQ circuit design and analysis process.

8.4   Conclusions

A simplified closed-form model of a three-terminal SFT device is described and implemented in Verilog-A. Although the proposed model does not include all possible superconductor-ferromagnetic multilayer interactions and proximity effects, the accuracy of the model is sufficient to support the SFQ-SFT circuit design process. The proposed model can be used to evaluate prospective SFT-based superconductive digital circuits and memory arrays, and requires low computational overhead, comparable to the standard RCSJ model of a Josephson junction.

Chapter 9
Inductive Coupling Noise in Multilayer Superconductive ICs

Large scale RSFQ circuits primarily use PTL interconnect for routing signals among standard cells [107, 146]. This type of interconnect, described in Sect. 4.1, consists of driver/receiver circuits and a stripline or microstripline [151]. This interconnect topology poses unique challenges on the routing process, as the available routing resources are severely limited. Different alternative topologies have been considered to reduce the number of metal layers required by these structures [248, 341]. These topologies, however, increase the inductive noise between adjacent striplines as compared to topologies with additional ground planes.

Another source of undesirable inductive coupling is the bias lines [50]. Bias current distribution networks are connected to each cell [49]. Despite the relatively small mutual inductance, high current within these bias lines can also couple to other circuits and transmission lines [51, 110]. Superconductive circuits are highly sensitive to inductance variations and parasitic coupling, further exacerbating this coupling issue.

In this chapter, issues related to inductive noise coupling in multilayer super-conductive ICs are discussed, and approaches to characterize and mitigate inductive noise in these systems are described [46]. Sources of coupling noise are reviewed in Sect. 9.1. The magnitude of inductive coupling noise in common circuit structures is described in Sect. 9.2. The effects of this coupling noise on circuit behavior are discussed in Sect. 9.3. A summary of this chapter is provided in Sect. 9.4.

9.1 Sources of Inductive Noise Coupling

In complex multilayer ICs, many sources of inductive coupling exist. PTL striplines (discussed in Chap. 3) are the primary type of transmission line for signal routing in VLSI RSFQ circuits [107]. The SFQ signals propagating along these lines can couple to other lines, producing noise. Metal lines carrying bias current can also

© Springer Nature Switzerland AG 2022
G. Krylov, E. G. Friedman, *Single Flux Quantum Integrated Circuit Design*,
https://doi.org/10.1007/978-3-030-76885-0_9

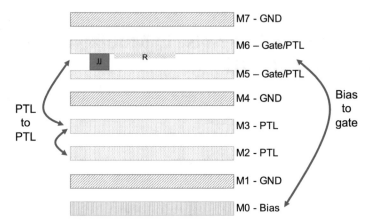

Fig. 9.1 Sources of inductive coupling noise in a superconductive IC (based on MIT LL SFQ5ee process)

couple to nearby circuits and PTLs, as shown in Fig. 9.1, affecting the operation of these circuits. Coupling noise can produce erroneous switching as well as reduce parameter margins [54]. In this section, these noise sources are described.

9.1.1 PTL Noise Coupling

In modern superconductive circuit fabrication, the metal resources are severely limited [139]. As discussed in Chap. 6, state-of-the-art niobium fabrication facilities only provide up to ten niobium layers, where the JJs are between specific layers. Several of these layers are required for the Josephson junctions and related connections, further reducing the number of available layers for routing [248]. A stripline within a PTL ideally consists of a signal line sandwiched between two ground planes. This structure is however impractical for large scale circuits.

Two approaches exist to reduce the number of layers used by PTLs, as shown in Fig. 9.2. One approach is to use a shared ground plane between two PTLs [341], as depicted in Fig. 9.2a. With this technique, five metal layers are necessary to produce two routing layers. Another approach is to utilize an asymmetric stripline, where two signal layers are sandwiched between two ground planes [248], as shown in Fig. 9.2b. For each signal routing layer in this structure, the distance to the ground planes and therefore the thickness of the dielectric layers are different. Moreover, no ground plane exists between the signal layers. This structure only requires four metal layers to produce two routing layers.

These approaches increase the mutual inductive coupling between the striplines. A tradeoff exists between the pitch of the striplines and the mutual inductance. A smaller pitch increases the available routing resources while also increasing the

Fig. 9.2 Topologies to allocate metal layers for two PTL routing strategies, (a) three ground planes [341], and (b) two ground planes [248]

noise coupling and decreasing the parameter margins. It is therefore important to characterize this tradeoff to determine the optimal pitch of the PTL tracks.

9.1.2 Coupling of Bias Current

The bias lines are typically located along the edge of the metal stack to reduce coupling of the bias current to sensitive RSFQ gates. Despite this approach, inductive noise coupling still occurs in these structures. High current flowing into the bias lines can affect the operation of the gates despite a small mutual inductance [48]. In addition, in certain topologies of the bias tracks, the bias lines are located close to the PTLs. This proximity can also introduce noise into the PTLs.

9.1.3 Techniques for Coupling Evaluation

To evaluate the magnitude of the inductive coupling, inductance extraction tools are necessary. Several field solvers are used to extract the coupling characteristics of these structures, as previously discussed in Sect. 7.4. FastHenry [285] is a popular and relatively accurate open-source tool for inductance extraction of superconductive circuits. The Sonnet field solver [277] is also popular, with additional features as compared to FastHenry. The results produced by these simulation tools can be used as guidelines for automated routing algorithms, as well as for the design of cell libraries. In the following sections, the magnitude and effects of inductive noise coupling are characterized to determine the minimum PTL track pitch and other guidelines for a variety of circuit structures and topologies.

9.2   Inductive Coupling for Common Circuit Structures

Although the topology of an IC layout is in general limited only by the design rules of the fabrication process, in a standard cell-based EDA flow, specific structures are common or ubiquitous. In this section, the coupling characteristics of these common circuit structures are described (based on FastHenry simulations).

9.2.1   Existing Experimental Data

To verify the correctness and relative error of inductance extraction in FastHenry, a comparison to other tools and/or experimental data is necessary. In this section, both the self-inductance and mutual inductance are compared to published experimental results.

The self-inductance of different structures in the MIT LL SFQ5ee process [139] is relatively well characterized as part of the design rules. A measurement of the self-inductance in FastHenry is performed by attaching a port to the input signal and ground planes while shorting the output signal line to ground. The self-inductance of different structures, extracted in FastHenry, is compared to existing experimental data. These self-inductances are in relatively close agreement—the error is on the order of 5%, sufficient for noise coupling evaluation.

The available data describing the mutual inductance within the SFQ5ee process is more scarce. The mutual inductance has been experimentally measured and published as part of the InductEx tool calibration process [342]. Mutual inductance extraction in FastHenry is performed in a similar way to the self-inductance—an additional port is attached to the second inductor. In Table 9.1, the experimental mutual inductance for these available data is compared to values extracted from FastHenry. These inductances are within 5%, exhibiting sufficient agreement for noise coupling analysis.

Table 9.1 Comparison of mutual inductance extracted in FastHenry with experimental data. The experimental data and layout topology are based on [342]

Layers	Experimental M, pH	FastHenry M, pH	Difference	Experimental standard deviation
M0-M1-M2-M7	3.37	3.58	+6.2%	1.12%
M1-M2-M3-M7	3.27	3.30	+0.9%	0.90%
M2-M3-M4-M7	3.04	3.13	+3.0%	2.04%
M3-M4-M5-M7	2.75	2.94	+6.9%	2.89%
M4-M5-M6-M7	1.95	1.89	−3.2%	1.43%

9.2.2 Coupling Between Parallel PTLs

In this section and the following sections, the PTL topology shown in Fig. 9.2b is evaluated assuming 5.2-μm-wide PTL signal lines and the ground planes in M1, M4, and M7. Structures with extended parallel PTLs in close proximity generally exhibit the largest inductive coupling coefficient, on the order of 10^{-2} for narrowly spaced lines. The dependence of this coupling coefficient on the separation between the signal lines is shown in Fig. 9.3 for coupling between two M1-M2-M4 PTLs or two M1-M3-M4 PTLs. These PTLs share the same ground planes, M1 and M4. The coupling coefficient exponentially depends upon the separation between the signal lines.

For PTL signal lines in adjacent layers sharing the same ground planes, the dependence of the coupling on the separation between layers is similar. For the topology depicted in Fig. 9.4 (PTLs in M1-M2-M4 and M1-M3-M4), this

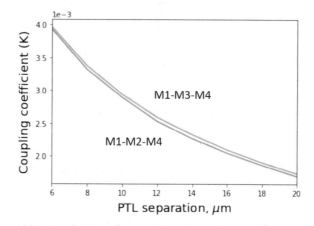

Fig. 9.3 Coupling between two identical parallel PTLs within the same layers with signal lines in M2 and M3

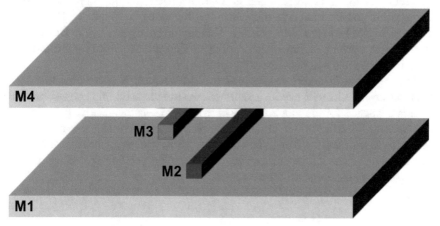

Fig. 9.4 Two identical parallel PTLs in adjacent layers with signal lines in M2 and M3

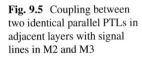

Fig. 9.5 Coupling between two identical parallel PTLs in adjacent layers with signal lines in M2 and M3

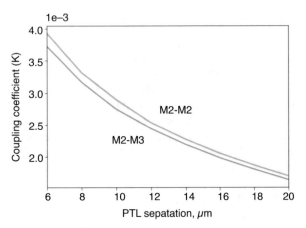

dependence is shown in Fig. 9.5. The vertical separation between these lines slightly reduces the coupling coefficient.

Coupling between the PTLs separated by a ground plane, e.g., M1-M3-M4 and M4-M5-M7, is significantly smaller, by approximately an order of magnitude. Any stitching vias—vias connecting the ground planes of a stripline (e.g., M1 and M4 in Fig. 9.5)—further reduce the inductive coupling. Any vertical overlap between signal lines in adjacent layers (shown in Fig. 9.7) drastically increases the coupling coefficient, proportional to the area of the overlap (or crossover).

9.2.3   Coupling Between Perpendicular PTLs With and Without Overlap

To reduce coupling between PTLs in adjacent layers, the interconnect tracks in these layers can be routed in perpendicular directions, as shown in Fig. 9.6 for PTLs in M1-M2-M4 and M1-M3-M4. This topology negates any practical effect from adjacent layer coupling—the coupling coefficient is reduced by two orders of magnitude as compared to the parallel case.

In a layout with highly constrained metal resources, it is possible that some adjacent layer PTLs will require a short overlap to reduce routing congestion. This case is evaluated for the topology shown in Fig. 9.7—perpendicular PTLs in M1-M2-M4 and M1-M3-M4—in Fig. 9.8. The zero on the horizontal axis corresponds to a simple crossing of perpendicular lines, while the larger offset corresponds to the length of the additional overlap. The coupling coefficient exhibits a linear dependence on the length of the overlap.

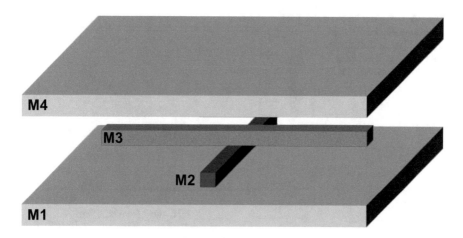

Fig. 9.6 Two perpendicular PTLs in adjacent layers with signal lines in M2 and M3

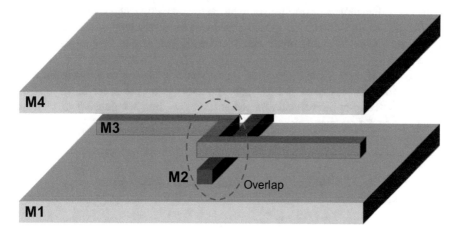

Fig. 9.7 Two perpendicular PTLs in adjacent layers with signal lines in M2 and M3 with a short overlap

9.2.4 Coupling Between M0 Bias Lines and Logic Gates

In the topology shown in Fig. 9.2b, the bias lines are routed in M0. Unlike PTLs that carry small voltage waveforms, these bias lines carry relatively high current, which can couple to the inductors within the sensitive RSFQ gates (e.g., in M5 and M6). This coupling is however reduced by the presence of two ground planes—M1 and M4—between the gates and bias lines.

Parasitic coupling to the RSFQ gate inductors depends upon the shape and relative position of the inductors. A critical practical case is assumed here—a long ($\sim 22~\mu$m) and narrow ($\sim 0.5~\mu$m) straight inductor in M5 with an inductance

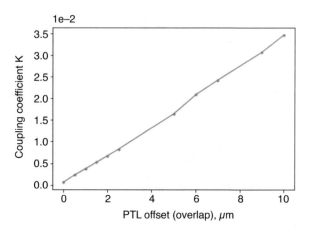

Fig. 9.8 Coupling between two perpendicular PTLs in adjacent layers with signal lines in M2 and M3 with a short overlap (see Fig. 9.7)

of ~ 8.5 pH (a typical inductance of a storage loop within a DFF). Cell libraries typically utilize more compact inductor geometries which exhibit a smaller coupling coefficient. A wider and longer inductor produces a higher coupling coefficient. This structure is however highly inefficient in terms of gate area and therefore unlikely to be used.

From FastHenry simulations, the approximate inductive coupling coefficient between the M0 bias lines and M5 gate inductors is on the order of 10^{-4} if the inductors overlap in the vertical dimension. This coupling is reduced to 10^{-5} if the inductors do not overlap.

9.3   Effects of Coupling on Circuits and Mitigation Guidelines

Different circuits exhibit a different sensitivity to parasitic coupling, thereby producing several types of errors. In this section, the effects of coupling on circuit operation are described.

9.3.1   Effects of PTL Noise Coupling

An SFQ pulse traveling on an active (aggressor) PTL produces a transient current spike at the receiver of the passive (victim) line, as shown in Fig. 9.9. Two cases are considered here, a small (~ 0.01) coupling coefficient and a large (~ 0.4) coupling coefficient. The case of a large coupling coefficient corresponds to PTLs in adjacent layers overlapping over a large distance. The case of a small coupling coefficient corresponds to all other practical routing topologies.

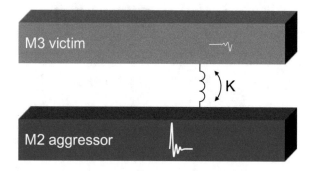

Fig. 9.9 Inductive coupling between parallel PTLs showing the aggressor and victim lines

The effects of PTL coupling noise are evaluated in WRspice [266]. WRspice does not normally support coupled transmission lines. To include coupling effects, a decoupling technique for lossless transmission lines is used [343–347].

In the case of small coupling, any additional current at the receiver of the victim line is on the order of a few μA—negligible as compared to the bias current of a typical PTL receiver. In this case, noise coupling momentarily degrades the bias margins of the receiver when this parasitic current waveform is present. The reduction in margins does not exceed a few percent.

In the case of large coupling, the additional current at the receiver of the victim line can exceed tens to hundreds of μA. This large current is typically not sufficient to switch the receiver junction of the victim line. If this coupled noise waveform however coincides with the signal waveform on the victim line, any additional current can prevent the receiver junction from switching. Unless the data signals on the aggressor and victim lines are synchronized to different phases of the same clock signal, this condition can eventually produce an error. To mitigate the effects of PTL coupling noise, the layout algorithm should avoid long overlapping PTLs in adjacent layers.

9.3.2 Effects of Bias Current Coupling

Coupling of bias current from the M0 lines to the gate inductors produces a different effect than transient PTL noise. These bias lines produce an additional constant current within the inductive loops of the gates.

The effect of bias current coupling on an RSFQ DFF is shown in Fig. 9.10. This additional current degrades the bias margins of the individual JJs within a gate, and therefore the overall bias margins of the gate.

To mitigate the effects of bias coupling on the gates, the layout algorithms should avoid vertical overlaps between the bias lines and the gate inductors. Alternatively, the bias current carried by the bias lines should be limited based on the maximum allowed change in the gate currents.

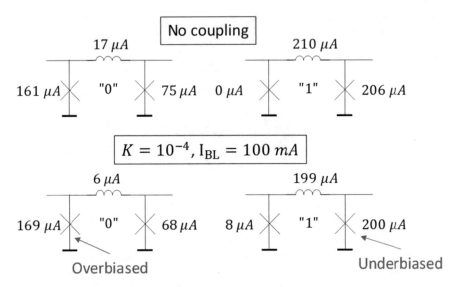

Fig. 9.10 Effects of bias coupling from the M0 bias line on current distribution within a DFF in M5-M6. I_{BL} is the current carried by the bias line. The stored state is shown in the middle of the storage loop

9.4   Conclusions

Inductive noise coupling is a primary issue in large scale ICs. Inductive coupling can reduce performance, introduce errors, and degrade parameter margins in both conventional and superconductive ICs. RSFQ circuits are particularly vulnerable to inductive noise due to the highly sensitive gates and low signal amplitudes. In this chapter, different sources of inductive noise in multilayer standard cell-based RSFQ circuits are reviewed. The coefficient of inductive coupling for the MIT LL SFQ5ee process is presented for different common circuit structures. Based on the magnitude of the coupling, the effects of inductive noise on the RSFQ gates and PTLs are described. Mitigation techniques are presented to reduce the effects of inductive noise on circuit behavior, enabling higher density SFQ circuits with wider parameter margins.

Chapter 10
Sense Amplifier for Spin-Based Cryogenic Memory Cell

One of the primary drawbacks limiting the use of superconductive electronics is the lack of a fast and dense memory capable of operating within a cryogenic environment [45]. Recent research suggests the use of cryogenic spin-based memory—magnetic tunnel junctions (MTJ) and spin valves (SV) [109]. In this chapter, a sense amplifier topology for a spin-based cryogenic memory cell is proposed and described [47].

The nanocryotron (nTron) device [84], previously described in Sect. 2.5.3, is used as a driver for a spin-based memory element—the cryogenic orthogonal spin transfer (COST) device [129], previously described in Sect. 3.6.3. A clocked DC-to-SFQ converter is used as a sense amplifier to resolve small differences in the readout current. The sense amplifier produces a variable number of SFQ pulses to represent different analog states by passing or blocking input clock pulses [54]. This clock is derived from the system clock by synchronizing the read pulse to the same clock signal. These output pulses are counted and converted into a binary form. The sense amplifier exploits the specific shape of the nTron output waveform characterized by an L/R time constant to achieve the resolution of low magnetoresistance (MR) memory cells and is adaptable to different nTron sizes, bias currents, and spin-based devices. The dynamic power dissipation and resolution of the sense amplifier can be adjusted by the frequency of the applied clock signal, allowing the resolution to be reduced for high MR devices. The sense amplifier consists of two Josephson junctions, requiring little area, particularly in comparison to a standard nTron device, and can therefore be connected to each column of the memory array.

This chapter is organized as follows. In Sect. 10.1, the circuit components of the memory cell and the topology of the sense amplifier are described. Simulation results are presented in Sect. 10.2. In Sect. 10.3, some conclusions are offered.

© Springer Nature Switzerland AG 2022

G. Krylov, E. G. Friedman, *Single Flux Quantum Integrated Circuit Design*,
https://doi.org/10.1007/978-3-030-76885-0_10

10.1   Circuit Components

In this section, the circuit components of the memory cell are introduced, and the topology of the sense amplifier is described. In Sect. 10.1.1, a readout scheme for a spin-based cryogenic memory cell is introduced. In Sect. 10.1.2, an SFQ flash A/D converter and a clocked DC-to-SFQ converter are presented, the advantages and disadvantages of these converters are discussed, and a clocked DC-to-SFQ converter is proposed for the sense amplifier. In Sect. 10.1.3, several connection topologies for the proposed amplifier are discussed.

10.1.1   Memory Cell Readout

For the purpose of the readout circuit, the memory cell is composed of a driver device and a spin-based memory element, such as an MTJ or spin valve device. The driver device considered here is the nTron [84], previously described in Sect. 2.5.3. The COST memory element [129], previously described in Sect. 3.6.3, is modeled as a variable resistive load of 5 or 5.5 Ω (corresponding to a magnetoresistance of 10%). Although the COST device exhibits a nonlinear dV/dI dependence on the current flowing through the device [129], it is negligible due to the magnitude of the currents used for readout ($\sim$45 μA). The memory cell is schematically shown in Fig. 10.1. The read enable pulse V_{read} switches the driver (nTron), diverting the bias current I_{bias} into the load (spin valve), exhibiting a resistance R_{load}.

The nTron device is switched on (to a resistive state) by an SFQ pulse (the read enable signal V_{read}). A resistive-inductive current divider is formed between the nTron device and the spin valve memory cell R_{load}. The current I_{out} that flows through the memory cell biases the sense junction J_2 within the sense amplifier. The magnitude and shape of the output current waveform are dependent on the resistance R_{load}, and therefore the logic state of the memory cell. The output waveform and relaxation time of an nTron, similar to an SNSPD (see Sect. 2.5.1), are characterized by an L/R time constant [348]. For readout, the nTron output current waveform is converted by the sense amplifier into a binary SFQ signal, later processed by the SFQ logic.

10.1.2   Synchronous DC/SFQ Converter as Memory Sense Amplifier

Due to natural quantization of magnetic flux, SFQ technology provides numerous benefits for efficient and fast analog-to-digital conversion [35]. Conventional SFQ analog-to-digital flash converters capable of resolving small differences in the nTron output current caused by a small MR require significant area for the resistive ladder

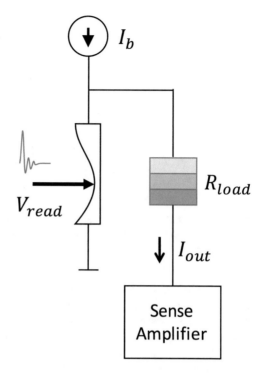

Fig. 10.1 Memory cell readout circuit. V_{read} is the read enable SFQ pulse, I_b is the bias current for the readout circuit, R_{load} is a spin valve element exhibiting a variable resistance, and I_{out} is the output current of the readout circuit

and inductive quantizer. In addition, these converters typically require higher current than provided by an nTron device of reasonable size. A sense amplifier suitable for analog-to-digital conversion within the memory array, requiring significantly smaller area, is therefore presented in this subsection.

The proposed sense amplifier is based on a well known timed DC/SFQ converter, described in [23]. The clocked DC-to-SFQ converter is schematically shown in Fig. 10.2. The converter consists of two JJs—a sense junction J_2 and an escape junction J_1. The incoming SFQ clock pulses are either transmitted by switching J_2 or blocked by switching J_1, depending upon the bias current of the sense junction. This bias current (I_{in} in Fig. 10.2 and I_{out} in Fig. 10.1) is the input of the converter. The two states of the memory cell produce two different resistances (R_{load}): therefore, two distinct L/R time constants and two different current waveforms at the input of the converter. The output of this sense amplifier is therefore in the form of a specific number of SFQ pulses which corresponds to the logic state of the memory cell.

As the state of the memory cell is distinguished by the number of SFQ pulses, this output needs to be converted into standard RSFQ convention (logic "one" is the presence of an SFQ pulse; logic "zero" is the absence of an SFQ pulse). This conversion can be achieved by a counter attached to the output of the readout circuit. The counter introduces significant overhead; however, only one counter per memory column is necessary. For prospective devices with improved MR, the number of generated SFQ pulses and therefore the size of the output counters can be decreased.

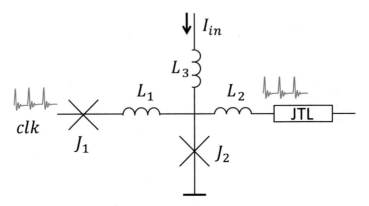

Fig. 10.2 Clocked DC-to-SFQ converter. J_1 and J_2 make up the decision-making pair of Josephson junctions. The SFQ clock signal passes through the circuit and is modulated by the input current I_{in}, which functions as a bias current for J_2

The operation of the sense amplifier for those devices with low MR can be enhanced by increasing the frequency of the input SFQ clock signal. This method enhances the resolution and speed of the amplifier but requires larger output counters to discriminate between the two memory states and increases complexity of the clock network [48].

The circuit described here is broadly used as a sensor circuit, including the previously mentioned hybrid Josephson-CMOS memory [349]. A few important distinctions exist with the previous work. This circuit is typically used to discriminate between two distinctly different current levels with a square-shaped input waveform. Sensing is therefore reliable and binary (the input current is either "on" or "off"). The proposed topology utilizes the exponential L/R decay of the input current to produce a different number of SFQ pulses. This approach therefore exploits the shape of the nTron output waveform while enabling the use of a well known, compact, and robust readout circuit.

10.1.3 Sense Amplifier Topologies

Several possible methods exist to connect the clocked DC/SFQ converter to the memory readout circuit. These methods are described in the following subsections.

10.1.3.1 Direct Connection

The output of the nTron-COST device is directly connected to the input terminal of the SFQ-to-DC converter without any additional bias. In this way, the overhead of the additional bias lines is avoided [50]. The sense junction (J_2 in Fig. 10.2) is, however, only biased by the input current. This method requires higher nTron output current, resulting in a significantly larger nTron device and bias current [51].

10.1.3.2   Additional Current Bias

An additional current source is connected to the input of the sense amplifier. This approach enables a smaller nTron for the readout circuit.

10.1.3.3   Additional Flux Bias

Additional flux is inductively coupled into the sense amplifier SQUID loop ($J_1 - J_2$ in Fig. 10.2). In this way, the benefits of additional bias current are combined with a more compact layout and lower energy dissipation.

10.2   Simulation Results

The proposed readout scheme is simulated in WRspice [268] for the low MR (10%) and high MR (100%) cases and two different SFQ clock frequencies, 20 and 50 GHz. All simulations in this section use the topology described in Sect. 10.1.3.1.

The circuit simulations utilize an RCSJ model of a Josephson junction with the fabrication process parameters for the MIT LL SFQ5ee process, and an unpublished compact model of the nanocryotron, which includes a nonlinear kinetic inductance and hotspot formation and relaxation [350, 351]. A compact model for a spin valve is not used since these devices are currently under development. Reliable switching of the MR states is assumed, and the spin valve element is modeled as two different resistances.

The component parameters for the simulations discussed in this section are chosen somewhat arbitrarily. The specific parameters are based on optimizing for a particular spin valve resistance and MR. The critical current of J_1 and J_2 is 125 μA. J_1 is critically damped ($\beta = 1$), while J_2 is slightly underdamped ($\beta \approx 3$). The connecting inductors L_1 and L_2 between the junctions are 2 pH. The current input is connected through L_3, a 1 pH inductor. The nTron driver of the memory cell includes a 12 nm gate, a 2 μm source/drain, and a 1.08 μm channel. The bias current of the nTron is varied and is mentioned in the caption for Figs. 10.3, 10.4, and 10.5.

The simulations are initiated with a read enable pulse applied to the gate of the nTron. This pulse switches the nTron, diverting the bias current into the memory element R_{load}. Different output waveforms are produced for different R_{load}. The bias current applied to J_2 within the sense amplifier therefore also varies. This condition produces a varying number of clock signals passed to the output of the sense amplifier, corresponding to the state of the memory element.

Waveforms for the low MR case are shown in Fig. 10.3. The resistance R_{load} of the memory cell is varied between 5 and 5.5 Ω, corresponding to 10% MR. The sense amplifier generates two or three SFQ pulses depending upon the state.

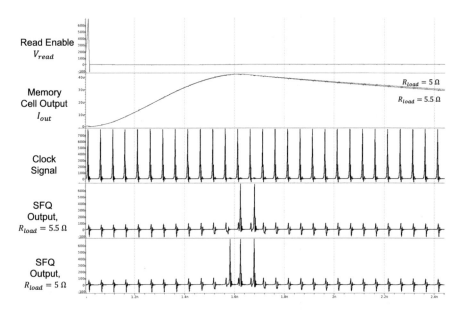

Fig. 10.3 Low MR waveforms. The clock frequency is 20 GHz. R_{load} is 5–5.5 Ω. The bias current is 74 µA. The sense amplifier output is two to three SFQ pulses

A larger output difference and greater robustness of operation are achieved for a memory element with a higher MR. Waveforms for the high MR case are shown in Fig. 10.4. The resistance of the memory cell is varied between 5 and 10 Ω, corresponding to 100% MR. The sense amplifier produces four or six SFQ pulses.

In these simulations, the clock frequency is set to 20 Ghz. The waveforms for the high MR case operating at a higher clock frequency (50 Ghz) are shown in Fig. 10.5. The sense amplifier produces 9 or 12 SFQ pulses depending upon the state. The synchronous DC-to-SFQ converter therefore greatly benefits from the higher clock frequency.

10.3　Conclusions

A sense amplifier to read memory cells composed of spin-based devices and nTrons is presented here. Guidelines for the sense amplifier are described, and simulation results for several MR values are shown. The sense amplifier is based on a widely used clocked DC-to-SFQ converter, which exploits the properties of the nTron output current. Enhanced speed and resolution for spin-based memory cells with a low MR of 10% make a clocked DC-to-SFQ converter a flexible and low area solution for reading out magnetic memory. A high MR combined with a high clock frequency results in robust discrimination of the memory states.

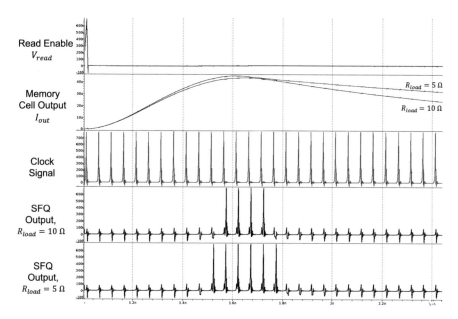

Fig. 10.4 High MR waveforms. The clock frequency is 20 GHz. R_{load} is 5–10 Ω. The bias current is 76.6 μA. The sense amplifier output is four to six SFQ pulses

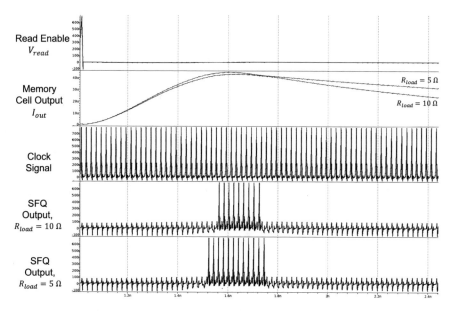

Fig. 10.5 High MR and high frequency waveforms. The clock frequency is 50 GHz. R_{load} is 5–10 Ω. The bias current is 76.6 μA. The sense amplifier output is 9–12 SFQ pulses

Chapter 11
Dynamic Single Flux Quantum Majority Gates

Among the major issues in modern large scale rapid single flux quantum circuits are the complexity of the clock network, tight timing tolerances, poor applicability of existing CMOS EDA techniques, and extremely deep pipelines which reduce the effective clock frequency. In this chapter, asynchronous dynamic SFQ majority gates are proposed to solve some of these problems. The proposed logic gates exhibit high bias margins and do not require significant area or a large number of Josephson junctions as compared to existing RSFQ logic gates. These gates exhibit a tradeoff among the input skew tolerance, clock frequency, and bias margins. Asynchronous logic gates greatly reduce the complexity of the clock network in large scale RSFQ circuits, thereby alleviating certain timing issues while reducing the required bias currents. Furthermore, asynchronous logic allows existing EDA tools to utilize CMOS approaches for synthesis, verification, and testability. The adoption of majority logic in complex RSFQ circuits also reduces the pipeline depth, enabling higher clock speeds in VLSI RSFQ circuits [48].

11.1 Introduction

Distribution of the clock signals is an important issue in VLSI RSFQ circuits [52]. Most RSFQ logic gates require a clock signal either to release the output pulse or to reinitialize the state of the gate to process the next datum [47]. A large scale circuit utilizing these gates requires a complex clock network. RSFQ circuits are capable of operating at extremely high clock frequencies (up to hundreds of gigahertz [34]), resulting in narrow timing tolerances. Multiple synchronization approaches exist for reducing the effects of timing on circuit operation, as previously described in Chap. 5. Most of these approaches, however, introduce significant performance, area, and bias current overhead as compared to conventional synchronous RSFQ circuits, partially negating the primary incentives of RSFQ technology.

© Springer Nature Switzerland AG 2022
G. Krylov, E. G. Friedman, *Single Flux Quantum Integrated Circuit Design*,
https://doi.org/10.1007/978-3-030-76885-0_11

Another distinctive feature which adversely affects the increasing integration of RSFQ circuits is limited fanout. Typical RSFQ logic gates and flip flops exhibit a fanout of one. A splitter gate is used to deliver a signal to multiple inputs [151]. This property necessitates large splitter trees that can dominate the area of a complex circuit. Insertion of these splitter trees into a logic path increases the delay of the path, significantly reducing the effective clock frequency of the overall system— degrading a primary advantage of RSFQ circuits. RSFQ clock networks consist mostly of splitters, decreasing the area available for logic and increasing the total bias current [51]. Novel solutions are therefore necessary to either support larger fanout or to reduce the number of splitters.

Existing RSFQ circuits typically utilize AND/OR/NOT logic with clocked AND and OR gates [153] and therefore exhibit these aforementioned issues. An alternative, functionally complete set is the MAJ/INV set, consisting of majority gates and inverters. Logic synthesis utilizing MAJ/INV logic has been shown to reduce logic depth, power, and delay in both CMOS benchmarks [297] and beyond CMOS technologies [352].

No efficient RSFQ majority gate has currently been described in the literature. A majority gate is typically composed of a combination of AND/OR/NOT gates and requires a clock signal with a network of multiple splitters. This topology therefore exhibits a large overhead both in area and delay and is therefore infeasible in complex circuits. In other SFQ logic families, such as quantum flux parametron (QFP), described in Sect. 3.5, and reciprocal quantum logic (RQL), described in Sect. 3.4, majority gates exist and are widely used [298, 353].

In this chapter, novel asynchronous RSFQ majority gates are described, based on the recently introduced dynamic SFQ (DSFQ) logic family [189]. In Sect. 11.2, a discussion of DSFQ is provided, and the benefits of DSFQ storage loops are discussed. Circuit design issues and related parameters for dynamic SFQ loops are discussed in Sect. 11.3. In Sect. 11.4, majority gates are introduced, the operation of these gates is described, and margin characteristics are presented. The application of majority gates to VLSI RSFQ circuits is discussed in Sect. 11.5. In Sect. 11.6, some conclusions are offered.

11.2   Dynamic SFQ Storage Loops

A novel family of asynchronous RSFQ logic gates—dynamic SFQ gates—is discussed in this section. Dynamic SFQ [189] is a recently introduced type of RSFQ logic, where the state of a gate is temporarily stored, and the gate self-resets to the initial state after a period of time. This capability enables asynchronous operation, significantly reducing the size and energy requirements of complex clock networks in large scale RSFQ circuits. Dynamic SFQ circuits are more similar to CMOS than regular RSFQ circuits, and can be separated into sequential and combinatorial logic [189], enabling the use of relevant CMOS EDA techniques.

The primary features of dynamic SFQ are self-resetting storage loops, as shown in Fig. 11.1. In conventional RSFQ, logic gates contain storage loops, which temporarily store a state based upon the input pulses received by the gate. The state of these storage loops is typically reset either by the clock pulse (e.g., DFF) or by another input pulse (e.g., Muller C element). In the absence of a reset pulse, logic gates can store a state indefinitely. This capability greatly complicates the application of certain CMOS techniques to RSFQ circuits and exacerbates flux trapping effects while increasing the complexity of the clock network.

Self-resetting storage loops have been used for many years and utilize either resistors or overdamped JJs inserted within the loops for flux leakage [354–356]. These loops exhibit a narrow timing tolerance due to a nearly constant rate of flux leakage, controlled by a single time constant. An example of this type of dynamic loop is shown in Fig. 11.1a. A recently introduced dynamic SFQ storage loop [189], shown in Fig. 11.1b, utilizes a critically damped JJ in parallel with a JJ in series with a resistor to reset the state. This topology produces two different time constants, for hold and reset, combining the slow leakage process during the hold period with fast relaxation to the initial state. A comparison of the current stored within a dynamic loop with single and double time constants, exhibiting a twofold increase in hold time, is shown in Fig. 11.2.

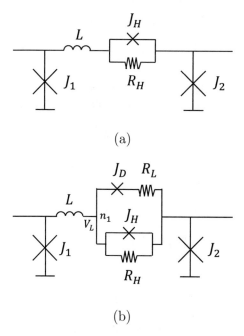

(a)

(b)

Fig. 11.1 Dynamic SFQ storage loops, (**a**) traditional dynamic loop, and (**b**) novel dynamic loop introduced in [189]

Fig. 11.2 Loop current (**a**) within a loop with a single time constant (dashed line) and (**b**) within a loop with dual time constants (solid line)

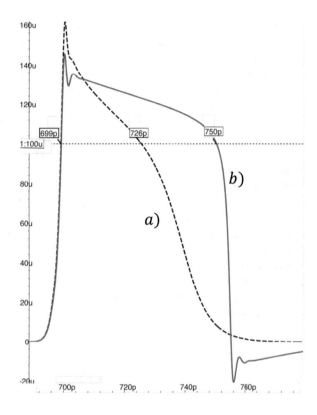

Novel DSFQ logic gates [189] combine the self-resetting storage loops with existing RSFQ circuits to achieve AND and OR logic functions. As the inversion operations can be propagated to the boundary of the combinatorial logic clouds [299] and integrated into registers with complementary outputs, these gates produce a functionally complete set.

In the following section, operation of dynamic SFQ storage loops is described. The effects of different circuit parameters are also discussed.

11.3 Circuit Design of Dynamic Loops

In this section, the operation and circuit components of DSFQ loops are described. The effects of these components on the operation of the logic gate are discussed, and guidelines for these parameters are provided.

Dynamic SFQ storage loops require a leakage mechanism for the magnetic flux to escape a superconductive loop. In the DSFQ loop shown in Fig. 11.1b, a combination of JJs and resistors is used. Initially, upon arrival of an input fluxon, the loop current is distributed between the two branches of the storage loop. The fraction of total current passing through resistor R_L produces a small voltage V_L ($\sim 15\ \mu V$) at the common node n_1. This current is dissipative and contributes to the initial slow leakage of flux. The voltage V_L produces a gradual increase in the phase of the junction J_H, setting the hold time. This increase in phase redistributes the loop current between the branches, eventually producing a 2π phase change in junction J_H, dissipating any remaining flux, and resetting the loop to the initial state.

Two major parameters characterizing a dynamic loop are the hold time τ_H and reset time τ_R. The hold time is set by the voltage V_L and is therefore affected by the resistance R_L and critical current $I_C(J_H)$ of junction J_H. A larger R_L and $I_C(J_H)$ reduce the fraction of current in the leaking branch, thereby increasing τ_H. The dependence of τ_H on R_L and J_H is shown in Fig. 11.3. While the same τ_H can be produced by different combinations of R_L and $I_C(J_H)$, a larger R_L for the same τ_H more quickly resets the gate to the initial state (with less current remaining within the loop after the same time), as shown in Fig. 11.4. Note that the reset time is faster with a larger series resistor R_L for a comparable hold time τ_H. A larger R_L is therefore generally preferable for dynamic loops.

The junction J_D serves two purposes in DSFQ circuits. The nonlinear inductance of J_D initially affects the distribution of current between the two branches of the dynamic loop. At the time of reset, this junction contributes to resetting the loop. The effect of the critical current $I_C(J_D)$ on τ_H and τ_R is depicted in Fig. 11.5. Although a larger $I_C(J_D)$ increases τ_H, it also increases τ_R at a faster rate. It is therefore preferable to use the smallest possible $I_C(J_D)$ to decrease the reset time.

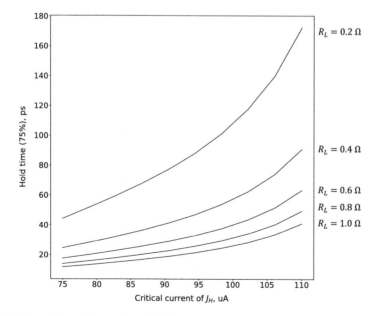

Fig. 11.3 Dependence of τ_H on R_L and $I_C(J_H)$

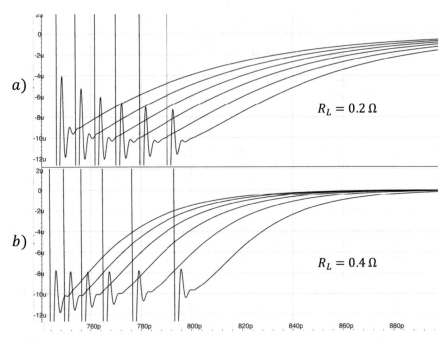

Fig. 11.4 Dependence of τ_R on R_L for similar τ_H ($\sim$ 45 to $\sim$ 90 ps), (**a**) $R_L = 0.2\ \Omega$, and (**b**) $R_L = 0.4\ \Omega$

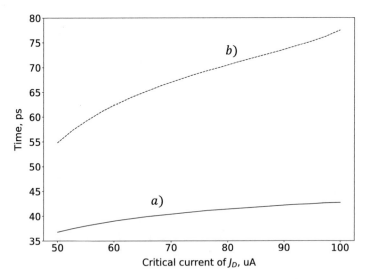

Fig. 11.5 Timing behavior of $I_C(J_D)$ in DSFQ loops, (**a**) dependence of τ_H on $I_C(J_D)$ (solid line), and (**b**) dependence of τ_R on $I_C(J_D)$ (dashed line)

11.4 Majority Gates

In this section, novel majority gates are proposed, and an example circuit configuration of a gate suitable for use in large scale RSFQ/DSFQ circuits is described. A margin analysis is also presented, and tradeoffs between the dynamic hold time and maximum clock frequency are discussed.

A proposed three-input DSFQ majority gate is shown in Fig. 11.6. The gate consists of three dynamic SFQ loops, as described in Sect. 11.3, $J_1 - J_4$, $J_2 - J_4$, and $J_3 - J_4$, with large storage inductors, L_A, L_B, and L_C. These loops share the common junction J_4. Individual loop currents combined with the bias current I_B contribute to the total current through J_4.

The proposed gate operates in a similar manner to the DSFQ AND gate [189]. Individual input pulses are temporarily stored in corresponding dynamic loops in the form of loop currents, and an output pulse is produced by J_4 upon the arrival of a second pulse within the hold time window τ_H. Individual loop currents decay shortly after this time, resetting the gate to the initial state. A third pulse, in the case of a "111" input, is stored for the same time τ_H. Other pulses cannot be accepted during this time, corresponding to the traditional CMOS setup time [101].

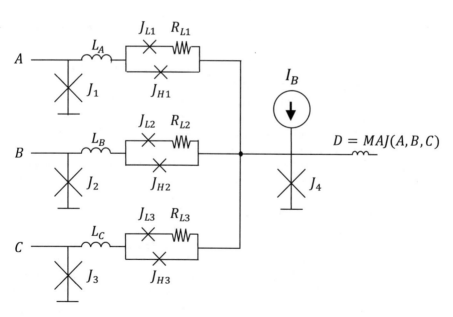

Fig. 11.6 Dynamic SFQ three-input majority gate. $I_C(J1, J2, J3) = 120\ \mu A$, critically damped. $L1, L2, L3 = 9.5$ pH (the input JJs and part of the inductance are shared with the input JTLs). $I_c(J_{L1}, J_{L2}, J_{L3}) = 50\ \mu A$, unshunted. $I_c(J_{H1}, J_{H2}, J_{H3}) = 106\ \mu A$, critically damped. $R_{L1}, R_{L2}, R_{L3} = 1.4\ \Omega$. $I_c(J_4) = 160\ \mu A$, critically damped. $I_B = 82\ \mu A$

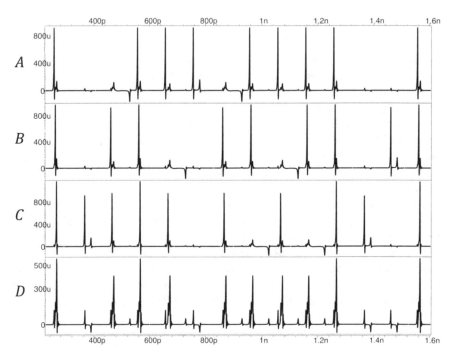

Fig. 11.7 Operation of the proposed three-input majority gate. A, B, C—inputs—and D, output

A simulation of the operation of the proposed gate is depicted in Fig. 11.7. The circuits are simulated in WRspice [268] based on the junction parameters of the MIT LL SFQ5ee process [139], with a JJ capacitance of 70 fF/μm. The clock frequency is 10 GHz, and the input data skew is 5 ps. Parameters of the dynamic loops are tuned to produce a 30 ps hold time—a maximum input skew for this parameter set. The average output delay of a three-input majority gate is 5.5 ps.

A layout of the proposed majority gate in the Hypres 4.5 kA/cm² technology [212] is depicted in Fig. 11.8. The parasitic inductances have been extracted using InductEx [286]. The primary effect of these parasitic inductances extracted from the layout is a small increase in the inductance of the corner loops as compared to the central loop. In addition, the inductance of the corner loops exhibits a small variation between the different branches. This variation changes the hold time between different branches of the gate. The largest variation in inductance for a three-input majority gate is 0.3 pH. For a more compact layout of the gate enabled by more advanced fabrication processes, this variation can be larger, reducing the bias margins. Wide conductors should therefore be used to connect the storage loops to J_4.

The simulated bias current margins for the proposed gate are ±20%. A higher (lower) bias current increases (decreases) the hold time, causing timing violations, which places limits on the upper and lower bias margins—a notable tradeoff for the

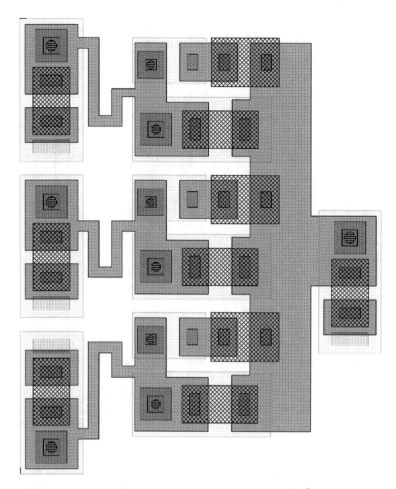

Fig. 11.8 Layout of a three-input majority gate in the Hypres 4.5 kA/cm^2 technology [212]

proposed gates. The delay and bias margins are for a manually optimized gate: these characteristics can be further enhanced.

Note that DSFQ gates also exhibit small variations in bias margins for different arrival times of the input pulses due to the dependence of the hold time on the bias current. In the simulations, all of the input pulses arrive within a 25 ps window, where the largest skew between individual inputs is 20 ps. For a 10 ps input skew, the bias margins are larger by 2 to 3%.

Another important tradeoff existing in the proposed gates is between the input skew tolerance and the maximum clock frequency. For high frequency operation, it is desirable to decrease the hold time, allowing the gates to reset faster to accept a new set of inputs. This choice reduces the tolerance of the gate to the input skew.

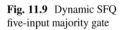

Fig. 11.9 Dynamic SFQ five-input majority gate

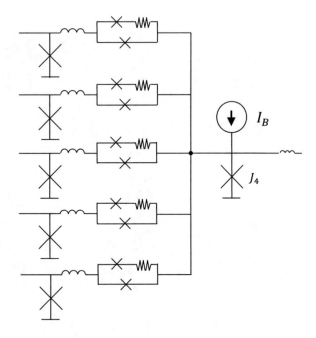

The frequency of operation can be increased if the input skew tolerance and target bias margins are relaxed.

It is similarly possible to increase the number of input loops for the proposed gate, producing a majority gate with five, seven, or more inputs. The five-input majority gate is schematically shown in Fig. 11.9. Operation of this gate is depicted in Fig. 11.10.

The extracted parasitic inductances of the five-input gate exhibit a similar small variation as in a three-input gate. This variation is due to the wide conductors connecting the dynamic storage loops to J_4 because of the large area of the gate. A more compact layout of a five (or more)-input gate in a more advanced technology is likely to exhibit greater variations in inductance as compared to a three-input gate, reducing the bias margins. These gates, although larger and less robust, increase the flexibility and benefits of majority gate logic.

11.5   Applications and Advantages

Possible applications for the proposed majority gates and majority-based RSFQ circuits are discussed in this section. The benefits of this approach and compatibility with energy efficient RSFQ are also discussed.

The primary purpose of the proposed gates is to enable majority-based large scale RSFQ circuits. Majority logic has been shown to reduce power and area in CMOS circuits and emerging beyond CMOS technologies [352]. Due to the need

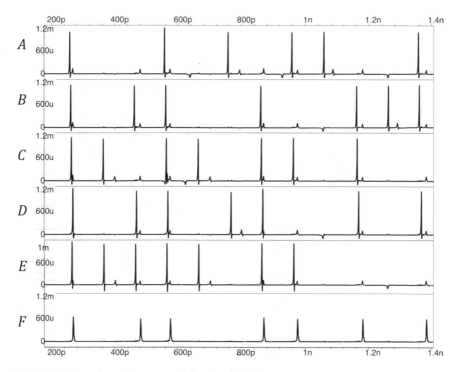

Fig. 11.10 Operation of the proposed five-input majority gate. A, B, C, D, E—inputs—and F, output

for splitter trees to support fanout, RSFQ technology is sensitive to a large logic depth, greatly increasing the delay and thereby lowering the frequency of operation in combinatorial logic. Majority logic circuits partially alleviate this issue, allowing the performance of large scale RSFQ circuits to be increased.

The reduction in logic depth is offset by the greater area and delay of the proposed gates, making this approach particularly suitable for specific classes of logic functions—in particular, nested expressions. Consider the following Boolean function:

$$Y = A * (B + C * (D + E * (F + G * H))). \qquad (11.1)$$

This function in conventional RSFQ AND/OR two-input logic consists of four AND gates and three OR gates and exhibits a logic depth of seven. A majority-inverter graph (MIG) representation [357] of this function with the same logic depth is shown in Fig. 11.11a. This majority-based expression can be optimized using the properties of majority logic [299], as follows:

$$Y = MAJ_3(A, 0, MAJ_3(B, 1, X)) \qquad (11.2)$$

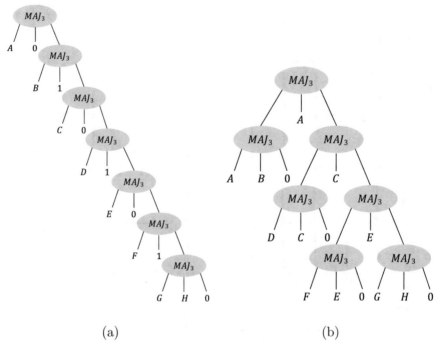

(a)   (b)

Fig. 11.11 Majority-inverter graph for (1), (**a**) original and (**b**) optimized

$$X = MAJ_3(C, 0, MAJ_3(D, 1, Z)) \tag{11.3}$$

$$Z = MAJ_3(E, 0, MAJ_3(F, 1, MAJ_3(G, 0, H))) \tag{11.4}$$

$$Z = MAJ_3(E, 0, MAJ_3(F, E, MAJ_3(G, 0, H)))$$
$$= MAJ_3(E, MAJ_3(F, E, 0), MAJ_3(G, H, 0)) \tag{11.5}$$

$$X = MAJ_3(C, 0, MAJ_3(D, C, Z))$$
$$= MAJ_3(MAJ_3(C, D, 0), C, Z) \tag{11.6}$$

$$Y = MAJ_3(A, 0, MAJ_3(B, A, X))$$
$$= MAJ_3(MAJ(A, B, 0), A, X). \tag{11.7}$$

The resulting MIG, depicted in Fig. 11.11b, is functionally equivalent to (11.1) and exhibits a logic depth of four. Specific degenerate majority gates in this graph, where one of the inputs is set to zero, can be mapped to regular DSFQ AND gates [189], further reducing the area.

Table 11.1 Comparison of different logic types for (11.1)

Logic type	# of gates	# of JJs	Logic depth	Minimal delay, ps	Additional cost
RSFQ [23]	7	80	7	56	7 splitters for clock tree (21 JJ)
DSFQ [189]	7	55	7	30.5	–
DSFQ majority logic	7	70	4	22	3 splitters for inputs (9 JJ)
DSFQ AND/MAJ	7	61	4	20	3 splitters for inputs (9 JJ)

A comparison of different circuits producing (11.1) is listed in Table 11.1. For RSFQ, the Stony Brook cell library [358] is used to estimate the number of JJs, while the delay characteristics are based on [249]. For DSFQ, a delay of 3.5 ps [189] is used for the AND gate, while a delay of the DSFQ OR gate is assumed to be 5.5 ps—equal to the delay of the confluence buffer from Figure 3 in [249].

For (11.1), majority-based DSFQ logic produces a smaller output delay (by 33%) due to the reduced logic depth with an area overhead of 25% as compared to regular DSFQ. By replacing degenerate majority gates with DSFQ AND gates, the area overhead is reduced to 10%, while the timing characteristics are further improved with an over 50% decrease in delay.

Conventional RSFQ logic gates are clocked. A multi-gigahertz clock network [326] is necessary to support operation of VLSI SFQ circuits, adding significant area for both the JJ and wiring layers [107, 146], dissipating higher power, and is a major source of added complexity for EDA tools. The proposed majority gates do not require a clock signal. Asynchronous combinatorial logic is therefore an attractive replacement for standard RSFQ logic.

Asynchronous RSFQ majority logic resembles conventional CMOS circuits by providing a boundary between the combinatorial logic and the sequential logic [305], facilitating the reuse of existing CMOS EDA methodologies and techniques [54, 154]. Tools and methodologies developed for majority-based emerging technologies, including QFP logic [359], can also be partially reused.

The proposed gates require reasonable area and number of JJs, similar to clocked RSFQ gates. The area can be further reduced by sharing the input JJs (J_1, J_2, J_3 in Fig. 11.6), and some of the storage inductances (L_1, L_2, L_3 in Fig. 11.6) with the preceding JTLs.

DFSQ majority logic, as DSFQ in general, is self-resetting. This feature supports the initial reset of the circuit and reduces some possible complications due to flux trapping. Flux trapped in holes or moats located close to the gates can couple to storage loops [360]. Self-resetting of these loops can mitigate this effect. While the flux trapped within the JJs will not be affected, this issue is not significant in modern small junctions [361].

The proposed gates are fully compatible with ERSFQ bias schemes [33]. The junction at the bias point (J_4 in Fig. 11.6) only switches once during a clock cycle. The average gate voltage therefore never exceeds the voltage of the bias bus produced by the feeding JTL [49, 50].

11.6   Conclusions

Novel asynchronous RSFQ majority gates are introduced based on the recently proposed dynamic SFQ logic topology. These gates utilize self-resetting storage loops to perform the majority function and reset to the initial state without requiring a clock signal. The proposed gates exhibit wide parameter margins and are capable of high frequency operation. Majority gates in large scale DSFQ-based RSFQ circuits reduce the pipeline depth, increase performance, and simplify the design process. The use of asynchronous logic gates greatly simplifies and reduces the size of the clock network and enables the use of certain CMOS EDA techniques by providing a well defined boundary between the combinatorial and sequential logic.

Chapter 12
Wave Pipelining in DSFQ Circuits

Dynamic SFQ (DSFQ) circuits, introduced in Chap. 11, are a promising type of asynchronous SFQ logic. The use of asynchronous logic lowers the complexity of the clock distribution network, greatly reducing power and area. The operation of DSFQ circuits significantly differs from both CMOS logic and conventional synchronous RSFQ logic. Novel design methodologies are necessary to synthesize DSFQ circuits, while increasing performance and decreasing area.

Path balancing, wave pipelining, and related timing methodologies to optimally synthesize DSFQ logic are described in this chapter. In Sect. 12.1, existing approaches to path balancing and wave pipelining in RSFQ circuits are reviewed. In Sect. 12.2, path delay balancing of DSFQ circuits is introduced. In Sect. 12.3, a methodology to enable wave pipelining in DSFQ circuits while reducing area is described. The chapter is concluded in Sect. 12.4.

12.1 Path Balancing and Wave Pipelining in RSFQ Systems

In conventional RSFQ circuits, most logic gates require a clock signal to operate. If the two inputs of a gate (e.g., gate F in Fig. 12.1) are connected to the gate outputs at different logic depths (for example, gates D and E in Fig. 12.1), an incorrect result will be produced. To prevent this error from occurring, path balancing (PB) elements (such as D flip flops (DFFs)) are inserted into the faster logic paths (such as from gate D to gate F in Fig. 12.1), as described in Sect. 7.5.2. These D flip flops do not perform any logical operation but delay the signal by a clock cycle to synchronize the data paths.

The number of path balancing DFFs required by a typical circuit is significant, frequently exceeding the number of logic gates [304]. Moreover, these flip flops require both a bias current and a clock signal, further increasing the area and complexity of the clock and power distribution networks [50, 51]. Path balancing is,

© Springer Nature Switzerland AG 2022

G. Krylov, E. G. Friedman, *Single Flux Quantum Integrated Circuit Design*,
https://doi.org/10.1007/978-3-030-76885-0_12

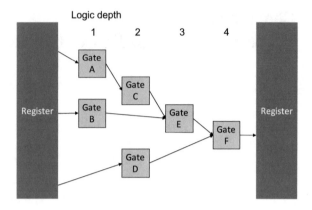

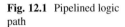

Fig. 12.1 Pipelined logic path

however, necessary to correctly operate RSFQ logic. Reducing the path balancing overhead during the synthesis process is an important design objective.

Wave pipelining [362] is a widely used technique in high performance circuits, where the pipeline stages operate at frequencies greater than the path delay should permit. The logic path between two registers is composed of two or more regions. Multiple data waves simultaneously propagate between these registers. This approach allows the next data wave to enter the combinatorial logic path before the previous data wave exits the same logic path, increasing the throughput and therefore the performance of the circuit.

Modern RSFQ circuits utilize wave pipelining in a limited fashion due to the large amount of clocked logic. Existing approaches utilize asynchronous AND/OR gates (or alternatively, clocked/resettable C elements) for computation, while handshaking signals propagate the data [52, 363]. The latency and throughput of these wave pipelined circuits are greatly improved as compared to circuits without wave pipelining. To initialize these intermediate asynchronous gates, a reset signal is required. The area overhead of wave pipelining utilizing resettable gates is comparable to the overhead of the clock distribution network. Self-resetting asynchronous RSFQ gates are an area efficient alternative; however, these gates, being a variation of a pulse merger [364], exhibit narrow timing margins.

Dynamic SFQ circuits [189] are a natural platform for RSFQ-based wave pipelining. The timing characteristics of DSFQ gates are adjustable over a wide range, and additional logic functions are available (e.g., majority gate) [48]. Similar to CMOS circuits, DSFQ circuits can greatly benefit from wave pipelining while avoiding the overhead of path balancing and clock distribution networks inherent to standard RSFQ circuits. To date, no path balancing or wave pipelining methodologies has been developed for DSFQ circuits. These methodologies and issues are the focus of this chapter.

12.2 Path Delay Balancing in DSFQ Circuits

Issues related to path delay balancing in DSFQ circuits are discussed in this section. In Sect. 12.2.1, the concept of delay balancing is introduced. In Sect. 12.2.2, a procedure to propagate inverters, necessary for correct operation of DSFQ circuits, is described. In Sect. 12.2.3, the benchmark circuits used to evaluate the proposed techniques are presented, and the RSFQ version of these circuits are characterized. In Sect. 12.2.1, the operation of DSFQ circuits without delay balancing is described.

12.2.1 Delay Balancing

The key characteristic of a DSFQ gate is the retention time—the time between the arrival of the last input pulse and the self-reset of the gate. The retention time affects the tolerance of a circuit to the input skew—the maximum skew between input signals of the same logic gate. An example of a DSFQ pipeline stage is schematically shown in Fig. 12.2. A stage is composed of multiple logic gates with different logic depths. The output of the gates at different logic depths (e.g., gate B and gate C in Fig. 12.2) exhibits a large difference in arrival time, which can exceed the input skew tolerance of gate D. A path balancing (or delay balancing) mechanism is therefore necessary for DSFQ circuits with significant input skews.

Since DSFQ-based combinatorial logic paths are not synchronized, inserting clocked DFFs is not a feasible approach for DSFQ. The delay elements are a

Fig. 12.2 Delay imbalance in DSFQ logic path. The pulse arrival time for the gates is highlighted. t_D and t_{int} are, respectively, the gate and interconnect delays

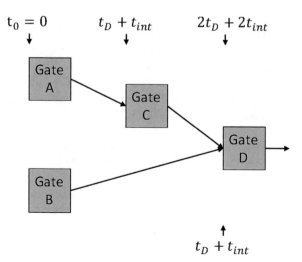

$t_0 = 0$ $\quad t_D + t_{int}$ $\quad 2t_D + 2t_{int}$

$t_D + t_{int}$

means to equalize the arrival time of the signals among different logic paths. A delay element does not perform any function apart from adding delay to the signals along a path. A Josephson transmission line with reduced bias current and increased damping can be used as an area efficient delay element [365]. A set of standard delay elements is necessary for automated path balancing during the logic synthesis process [304]. These delay elements exhibit tunable delays which correspond to the delay of the other elements within a cell library.

DSFQ gates exhibit a tradeoff between the input skew tolerance and system throughput, both affected by the retention time. For increased throughput, it is desirable to decrease the retention time and therefore tolerance to the input skew. The primary source of delay uncertainty—input skew in DSFQ circuits—is caused by the difference in fanout among different input signals.

SFQ signals require an active splitter gate to drive multiple nets [151]. The most commonly used splitter is a binary splitter with cascaded binary splitter trees for fanouts greater than two. Splitter gates introduce significant delay, large as compared to the delay of the interconnect, comparable to the delay of the logic gates [151]. Standard splitter gates suitable for routing typically use passive transmission lines (PTL) which increases the path delay since a PTL includes both a driver and receiver in each line [107, 146]. These features produce a significant difference in the delay between signals with differing fanout. If these signals are connected to a DSFQ gate, the gate needs to tolerate a large input skew to prevent errors. The effects of the splitters on delay balancing are further discussed in Sect. 12.2.4.

12.2.2   Inverter Propagation

An essential step in the synthesis of DSFQ circuits or conversion of RSFQ circuits into DSFQ is inverter propagation. DSFQ logic gates are asynchronous; however, no asynchronous inverter exists for DSFQ logic. This issue is alleviated by propagating standard inverters to the boundaries of the combinatorial logic, and replacing standard flip flops with flip flops with complementary outputs (e.g., DFFC [241]) or similar cells, as shown in Fig. 12.3a.

To propagate inverters, De Morgan's laws of Boolean logic are applied. The preceding AND gates are replaced with OR gates and vice versa, and the inverters are moved from the gate outputs to the gate inputs. Majority gates are not changed; only the inverters are moved.

A more complex case is when a gate exhibits multiple fanout, where both the inverted and non-inverted output signals are used by other gates. In this case, the gate is duplicated, maintaining the original non-inverting paths, while a new inverting path is introduced with appropriate logical modifications, as shown in Fig. 12.3b. While this redundant approach increases area by adding a duplicate gate, fewer splitters are required as compared to the original circuit. All of the inverters in DSFQ circuits are propagated to the boundary of the combinatorial logic.

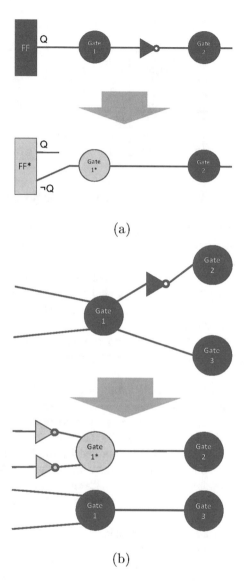

Fig. 12.3 Inverter propagation process, (**a**) at edge of the pipeline stage, and (**b**) complicated by the presence of splitters. The unchanged gates are depicted in the darker shading

12.2.3   Benchmark Circuits

DSFQ circuits utilizing different approaches to path balancing are compared to equivalent RSFQ circuits in the following sections. A subset of ISCAS'85 benchmark circuits is considered here for this evaluation. These circuits are synthesized

Table 12.1 Characteristics of ISCAS'85 benchmark circuits mapped to RSFQ cell library with path balancing

Benchmark circuit	Gates	Logic levels	Inputs	PB DFFs	Splitters	Clock splitters	Inverters	Latency (ps)	Total gates
c17	8	5	5	6	3	7	2	100	26
c432	369	27	36	634	140	368	60	540	1571
c880	434	22	60	743	155	433	76	440	1841
c1335	680	22	41	1284	251	679	72	440	2966
c2670	1120	22	233	1263	300	1119	163	440	3965

and mapped to a generic RSFQ/DSFQ cell library using the open-source synthesis tool, ABC [295].

These mapped RSFQ/DSFQ circuits are path balanced. In DSFQ circuits, the inverters are propagated to the input boundary of the logic and integrated with flip flops using DFF cells with complementary outputs. Although these gates are significantly larger than conventional DFFs, the total number of these flip flops does not exceed the number of circuit inputs, which corresponds to approximately 10% to 20% of the gates in a circuit, as listed in Table 12.1. The resulting Verilog netlists are combined with RSFQ and DSFQ SystemVerilog gate models and simulated using Synopsys VCS. The RSFQ circuits are used as a reference. The operation of the DSFQ circuits is compared to RSFQ to verify the correctness of the circuit modifications.

For both RSFQ and DSFQ, the number of gates, splitters (in both the logic paths and the clock distribution network), and inverters is determined for each of the benchmark circuits. The gates are assumed to be approximately equal in area. Although this assumption is not accurate, the total number of gates provides a high fidelity approximation of the total area required by the benchmark circuits. The timing characteristics of the gates are also simplified—the gate delay is assumed to be 8 ps for all gates, flip flops, and splitters with any fanout; while the interconnect delay is 6 ps between any gates. The characteristics of the RSFQ benchmark circuits are listed in Table 12.1.

Note that the overhead of the path balancing process in conventional RSFQ circuits is significant, often exceeding the area occupied by the logic gates. Clock splitters are another significant contributor to the overall circuit area. The latency of the RSFQ circuits is proportional to the logic depth and clock period. In Table 12.1, the clock period is assumed to be 20 ps (50 GHz).

12.2.4   DSFQ Circuits Without Path Balancing

Unlike RSFQ circuits, DSFQ circuits, under specific conditions, produce the correct output without path balancing. If the two inputs of a DSFQ AND/majority gate are on different logic levels, as shown in Fig. 12.4, the faster signal path (with shorter

Fig. 12.4 Delay imbalance in DSFQ logic path. Faster and slower signal paths are highlighted. The circuit operates correctly if the retention time is sufficient

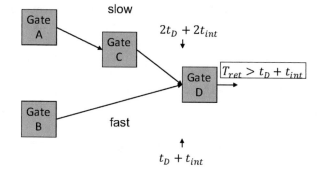

Table 12.2 Characteristics of ISCAS'85 benchmark circuits mapped to a DSFQ cell library with and without splitters

	Without splitters				With splitters				
Benchmark circuit	Gates	Logic levels	Maximum latency, ps	Maximum input skew, ps	Gates	Splitters	Logic levels	Maximum latency, ps	Maximum input skew, ps
c17	6	4	48	14	6	2	6	76	42
c432	309	26	356	126	309	128	36	496	350
c880	358	22	300	168	358	32	162	440	378
c1335	608	20	272	168	608	30	244	412	364
c2670	957	18	244	168	957	25	417	342	196

logic depth) changes the state of the gate, which persists within the gate during the retention time. If the slower signal path (with deeper logic depth) arrives during this time, a correct output is produced by the gate. In the case of a DSFQ OR gate, the correct output is produced immediately upon arrival of the first input pulse, while the gate remains insensitive to the second input pulse during the retention time. Therefore, if all of the gates within a DSFQ circuit exhibit a sufficiently long retention time, path balancing is not required to maintain correct operation.

A retention time that is larger than necessary does not affect the operation of a circuit if the following datum is not expected during this time. A larger retention time increases the robustness of the circuits to variations in the input skew. The retention time also exhibits a negligible effect on the area of the DSFQ circuits. Due to these properties, it is convenient to set the retention time equal to the maximum input skew for all DSFQ gates.

This approach is evaluated using benchmark circuits, as described in Sect. 12.2.3. The properties of the DSFQ circuits without path balancing are listed in Table 12.2. Observe that inclusion of the splitters greatly increases the maximum input skew of a circuit (sometimes by more than twofold). DSFQ circuit analysis which does not consider the splitters will produce incorrect results; the splitters should therefore not be neglected when evaluating the retention time or during the path balancing process.

As compared to the RSFQ circuits listed in Table 12.1, DSFQ circuits require significantly less area and generally exhibit smaller latency. Although DSFQ circuits can operate without path balancing, this approach results in a large input skew which necessitates a long retention time. The gates cannot accept another datum during this time, limiting the throughput. This limitation can be circumvented by using wave pipelining, as discussed in the following section.

12.3   Partial Path Balancing and Wave Pipelining

Approaches to enable wave pipelining in DSFQ circuits are discussed in this section. In Sect. 12.3.1, necessary conditions for wave pipelining are discussed. In Sect. 12.3.2, full path balancing (for all gates and splitters) is explored as an approach to wave pipelining. Partial path balancing is introduced in Sect. 12.3.3.

12.3.1   Necessary Conditions for Wave Pipelining

To enable wave pipelining in DSFQ circuits, any uncertainty in the arrival time of the data signal for each gate should be managed. The primary sources of timing uncertainty in DSFQ gates are the difference in logic depth of the input signals, and the difference in the number and fanout of the splitters along these input paths. If these sources of uncertainty are mitigated, the difference in the interconnect length remains as a source of input timing skew. This difference is typically small for a combinatorial circuit, as the gates are typically located in close proximity, and the interconnect delays are small due to the high propagation velocity of the SFQ pulses within PTLs.

In wave pipelined DSFQ circuits, the following data wave can be initiated if any collisions with the existing data waves along the signal path are avoided. A new data wave is launched into the signal path when the data wave already present in the pipeline is located far from the input (for example, near gates 5 or 6 in Fig. 12.5), avoiding any collision with the existing data wave. The minimum time (or data

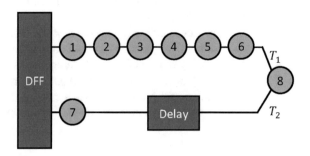

Fig. 12.5 DSFQ pipelined signal path supporting multiple data waves

Fig. 12.6 Full path balancing in DSFQ circuits. The difference in path delays related to the gates and splitters (shown as circles) in the slower (top) path is balanced in the faster (bottom) path by delay elements D

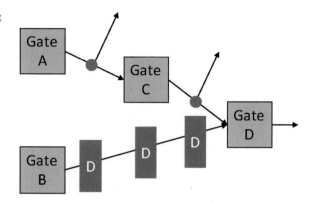

skew) before the following data wave can be applied to the pipelined signal path is the largest retention time of all of the gates within a signal path. The objective is therefore to reduce the maximum input skew of the DSFQ gates within a circuit.

12.3.2   Full Path Balancing

A straightforward approach to reduce the input skew is to utilize RSFQ-like path balancing. Any difference in the logic depth of the gate inputs is compensated by adding delay elements, with the delay equal to a typical gate delay or tuned to a specific delay within the slower path, as shown in Fig. 12.6. In addition, the delay of the splitters, as discussed in Sect. 12.2.4, can also be compensated by adding delay elements. Although some of the successive delay elements can be merged, this process can produce collisions between the data or limit overall throughput.

This approach is applied to the benchmark circuits discussed in Sect. 12.2.3. The results are listed in Table 12.3. Note that full path balancing greatly reduces the minimum data skew, limiting any differences in the input arrival time to the difference in interconnect delay. As each logic level is a separate pipeline stage, this approach produces the ultimate throughput of a DSFQ circuit. This approach, however, exhibits significant area overhead as compared to conventional RSFQ circuits. This significant area overhead makes full path balancing impractical in DSFQ circuits. In addition, the benefits of higher retention times in DSFQ circuits are not exploited.

12.3.3   Partial Path Balancing

Each gate in a pipeline stage has a different input skew depending upon the preceding gates and splitters. The majority of gates exhibits zero or small input

Table 12.3 Characteristics of ISCAS'85 benchmark circuits mapped to a DSFQ library with full path balancing

Benchmark circuit	Gates	Logic levels	Splitters	PB elements (logic)	PB elements (splitters)	Maximum latency, ps	Maximum input skew,[a] ps	Total gates	Total gates (compared to RSFQ)
c17	6	6	2	3	5	76	0	16	−40%
c432	309	36	128	2392	482	496	0	3311	+111%
c880	358	32	162	3001	743	440	0	4246	+132%
c1335	608	30	244	2744	1092	412	0	4688	+58%
c2670	957	25	417	4258	670	342	0	6302	+59%

[a] The input skew is due to the gates and splitters; any difference in interconnect delay also contributes to the input skew

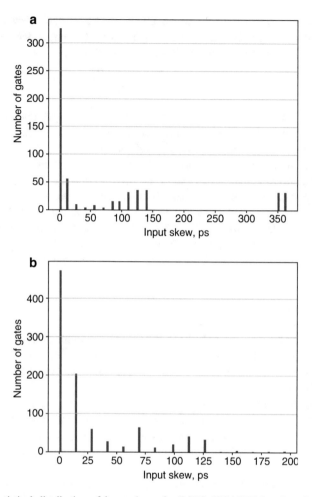

Fig. 12.7 Statistical distribution of input skews for DSFQ ISCAS'85 benchmark circuits, (**a**) c1355, and (**b**) c2670

skew. Certain critical paths, however, exhibit significant or close to maximum skew. A statistical distribution of the input skews is shown in Fig. 12.7 for two different benchmark circuits.

In a DSFQ circuit, delay balancing between paths with near zero or small skew does not increase performance while expending area. Delay balancing of the critical paths, however, reduces the maximum input skew of a circuit while increasing the throughput.

Partial delay balancing is the path balancing process for these critical paths. The distribution of the input skews is initially evaluated, and the target throughput is determined. The target throughput is the number of data waves simultaneously propagating within a logic path times the effective clock period. This number of simultaneously propagating data waves affects the target data skew T_{DATA}—the period of time between two consecutive data waves. The target data skew T_{DATA} sets the limit for the input skew within a circuit, as follows,

$$T_{DATA} \geq T_{ret} \geq T_{skew}^{max}, \qquad (12.1)$$

where T_{skew}^{max} is the maximum input skew, and T_{ret} is the gate retention time. By reducing the input skew for a small number of outlier gates, the data skew T_{DATA} can also be decreased, increasing overall performance.

Based on the target T_{DATA}, those gates with an input skew $T_{skew} \geq T_{DATA}$ are delay balanced—the delay elements are inserted into the faster path to reduce the gate input skew. To accommodate multiple data waves, a specific number of delay elements are inserted to avoid data collisions. For N data waves, $N - 1$ delay elements with a delay time of $T_d = T_{skew}/N$ are inserted into the faster path. A single datum is also stored in a DSFQ gate for $T_{ret} = T_d$. After completion of this process, the gate exhibits a small input skew and does not limit the overall data skew.

12.3.3.1   Case Study

Consider the DSFQ version of ISCAS'85 benchmark circuit c880 with splitters, as listed in Table 12.2. This circuit can operate without path balancing for data skew $T_{DATA} \geq 378$ ps. A statistical distribution of the gate input skews for this circuit is shown in Fig. 12.8a.

Based on this distribution and the target performance, T_{DATA} is arbitrarily set to 210 ps. Gates with a larger input skew are identified and path balancing is applied. The critical paths are eliminated from the circuit, and the new maximum input skew is now T_{DATA}. The distribution of input skews in the modified circuit is shown in Fig. 12.8b.

The resulting partially path balanced circuit can now operate at $T_{DATA} - 210$ ps. In this case, 128 standard delay elements (each element with a delay of 8 ps) are inserted—a significant improvement as compared to both RSFQ and DSFQ circuits with full path balancing. Moreover, the total additional delay is approximately 1.8 ns, where only 13 discrete delay elements are required to avoid collisions between data waves.

Partial delay balancing increases the performance of DSFQ circuits by enabling wave pipelining, while requiring low area overhead as compared to RSFQ and DSFQ circuits with full path balancing. These results also highlight the need for efficient compact delay elements. Only a few discrete delay elements are required for path balancing; the remaining delay elements could utilize a few elements with a large delay. To date, no compact circuits capable of producing a delay on the order of ~ 100 ps exist. Efficient delay elements would greatly reduce the area required for path balancing in DSFQ circuits, further incentivizing this SFQ circuit topology.

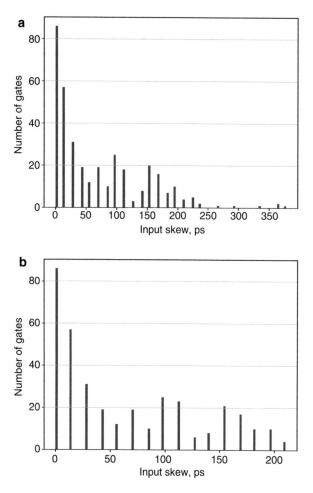

Fig. 12.8 Statistical distribution of input skews for DSFQ ISCAS'85 benchmark circuit c880, (**a**) before partial path balancing, and (**b**) after partial path balancing

12.4 Conclusions

DSFQ circuits can greatly reduce the area and complexity of the clock and bias distribution networks in large scale superconductive circuits. The path balancing process, essential for RSFQ circuits, is less important for DSFQ circuits. Path delay balancing can, however, be used to increase the performance of DSFQ circuits by enabling wave pipelining. In this chapter, different path balancing approaches for DSFQ circuits are evaluated and compared to equivalent RSFQ circuits. Full delay balancing produces the highest throughput by enabling wave pipelining while minimizing the data skew. This approach, however, exhibits an impractical area overhead and does not exploit the improved timing characteristics of DSFQ circuits. A partial path balancing methodology is described, where path balancing is primarily applied to the critical paths. This methodology enables wave pipelining in DSFQ circuits, reducing the period between data waves, while requiring relatively small area, thereby providing a tradeoff between area and throughput.

Chapter 13
Design Guidelines for ERSFQ Bias Networks

ERSFQ is an energy efficient, inductive bias scheme for RSFQ circuits, where the power dissipation is drastically lowered by eliminating the bias resistors while the cell library remains unchanged. Operation and components of the ERSFQ bias scheme are discussed in Sect. 4.3.2.4. An ERSFQ bias scheme requires the introduction of multiple circuit elements—current limiting Josephson junctions, bias inductors, and feeding Josephson transmission lines (FJTLs).

In this chapter, parameter guidelines and design techniques for ERSFQ circuits are presented. These guidelines enable more robust circuits resistant to severe variations in supplied bias current. Trends are considered and advantageous tradeoffs are discussed for the different components within a bias network. The guidelines provide a means to decrease the size of a FJTL, and thereby reduce the physical area, power dissipation, and overall bias current, supporting further increases in circuit complexity [51]. A distributed approach to ERSFQ FJTL is also presented to simplify placement and minimize the effects of the parasitic inductance of the bias lines. This methodology and related circuit techniques are applicable to automating the synthesis of bias networks to enable large scale ERSFQ circuits [49, 50, 166].

13.1 Introduction

A major obstacle to improving the large scale integration of ERSFQ circuits is the lack of electronic design automation (EDA) tools [269]. Current research is aimed at adapting existing CMOS-based industrial tools [283] while developing novel algorithms and circuit techniques specifically targeted for SFQ technology [366], as discussed in Chap. 7.

RSFQ gates are current biased and, unlike CMOS, require a precise bias current to maintain correct functionality. Both over- and underbiased gates can produce logic errors [154]. Proper distribution of the bias currents within SFQ circuits is

© Springer Nature Switzerland AG 2022
G. Krylov, E. G. Friedman, *Single Flux Quantum Integrated Circuit Design*,
https://doi.org/10.1007/978-3-030-76885-0_13

therefore critical for continuing the integration of SFQ circuits toward LSI and VLSI levels of complexity. As the bias lines in SFQ circuits are lossless and inductive, EDA tools require a novel set of guidelines, heuristics, and algorithms for the automated generation of bias networks for SFQ-based VLSI circuits. All of these issues and concerns emphasize the importance of correct and efficient distribution of the bias currents within large scale ERSFQ circuits. These bias distribution structures need to be synthesizeable by prospective SFQ EDA tools to support the increasing complexity of ERSFQ circuits.

Some commonly used ad hoc design approaches [161, 162] as well as guidelines [163–165] currently exist on the proper design of ERSFQ bias networks. One rule of thumb is related to the size of the feeding JTL, which is typically chosen to ensure the FJTL bias current is about 25% to 30% of the load bias current. The dependence of the margins on the operating frequency of a feeding JTL has also been considered [161].

The importance of the parasitic inductance of the bias bus is emphasized in [164], where different bias inductances produce different current distributions. This dependence is discussed in Sect. 13.3.5, where design guidelines for increased robustness of the bias network are provided. Techniques to reduce current deviations in ERSFQ bias networks have been proposed in [165], along with some guidelines on FJTL size. Another study has suggested optimal values for certain ERSFQ component parameters, such as the bias inductance and size of the FJTL [163]. The effect of the size of the FJTL on the operational range of the ERSFQ bias networks is discussed here, and design guidelines for a preferable FJTL size are provided. Among the issues first discussed here are the optimal topology of the FJTL stage and the effect of the FJTL bias margins on the circuit bias margins. As the size and topology of the feeding JTL affect the physical area and bias currents, additional guidelines are required to integrate ERSFQ circuits into an industrial design flow.

In this chapter, a semiautomated analysis methodology is used to develop novel parameter guidelines, as well as expand and clarify some existing guidelines. In addition, an ERSFQ topology utilizing multiple clock domains—an extension of the approach proposed in [164]—is described. The chapter is organized as follows. In Sect. 13.2, a semiautomated analysis methodology is presented. This methodology is used to develop design guidelines for ERSFQ bias networks, as described in Sect. 13.3. In Sect. 13.4, a distributed placement methodology for ERSFQ bias networks is presented. In Sect. 13.5, some conclusions are offered.

13.2 Example Circuit and Analysis Methodology

An ERSFQ bias network is composed of a variety of different elements, where each gate contains a highly nonlinear JJ as a current regulator. This structure makes infeasible the development of closed-form analytic expressions characterizing the behavior of a bias network [45]. An analysis of the bias network is therefore limited to observations of trends and the effects of different component parameters on circuit behavior.

For this analysis, a semiautomated script is used to perform multiple circuit simulations in the WRspice simulator [268] to extract behavioral trends. Two primary circuit components of an ERSFQ bias network are the load, which requires a bias current I_B, and a feeding JTL, which functions as a voltage source with a maximum average voltage V_B. This topology, schematically shown in Fig. 13.1, is used to monitor the bias current within the load and to extract parametric trends.

A general ERSFQ circuit is used as a standard load: specifically, a shift register composed of multiple D flip flops chained together with JTLs. A shift register, a common topology in complex ERSFQ circuits, is extendable and robust and exhibits wide parameter margins. The shift register is synchronized by an H-tree clock distribution network consisting of a binary splitter tree [151] buffered by JTLs. In this analysis, 16-stage, 32-stage, and 64-stage registers are used to manage the simulation time. An H-tree clock network is chosen due to wide parameter margins and to increase the size of the load.

A chain of JTL stages is used as a FJTL. The number of stages is varied, and the results presented in Sect. 13.3 are described in terms of the total FJTL bias current. The FJTL can ideally operate at the same clock frequency as the rest of the circuit. This approach dissipates minimal power while maintaining ERSFQ operation. The range and robustness of such an operation are, however, reduced. Higher FJTL frequencies, although beneficial for system bias margins, dissipate more power. In

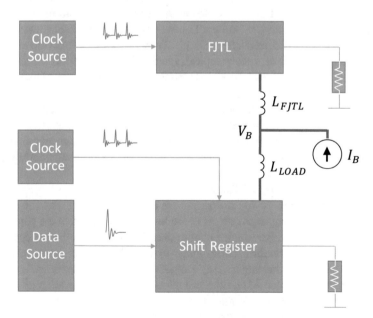

Fig. 13.1 Example topology of the ERSFQ-biased circuit used in this analysis

this analysis, the system clock frequency is 10 GHz, while the FJTL frequency is 12.5 GHz.

The data source generates a train of SFQ pulses with a specific pattern, representing the input data. This pattern affects the activity factor of the load, directly changing the gate voltage (V_{GATE} in Fig. 4.18), and therefore the behavior of the bias network. The clock sources generate a train of SFQ pulses representing the primary system clock signal. The output of both the chain of feeding JTL stages and the shift register is terminated by a resistive load. All circuits are from the Stony Brook cell library [241] and [23]. As the Stony Brook cell library utilizes normalized parameter values, the standard critical current of the JJs is 250 µA. The bias margins of the FJTL are varied, as described in Sect. 13.3.3, but are generally above ±40%. The bias margins of the shift register are above 30%.

The input and output data streams of the shift register are compared to verify circuit operation. The output of the WRspice simulation is parsed to extract all 2π phase transitions of the input and output JJs. These transitions are aligned with the transitions of the JJs within the clock network, where the clock transitions are measured at the sinks of the network. The resulting bit patterns are compared. This methodology supports a bias margin analysis, where the bias current is either swept linearly or modified in a binary search pattern, and the upper/lower bounds that ensure correct operation are determined.

The primary metric of robustness of operation in RSFQ circuits is the bias margins. The bias margins are a measure of the additional or absent bias current tolerated by a circuit. The ERSFQ bias networks affect the bias margins due to dynamic redistribution of the bias currents between the FJTL and the many loads. The ERSFQ bias margins are limited, however, by the intrinsic bias margins of a properly biased and optimized RSFQ circuit. These margins typically do not exceed 20% for circuits of intermediate complexity and are often lower for more complex circuits [135]. It is therefore infeasible to optimize an ERSFQ bias network within a large circuit to achieve margins of operation wider than 20% to 30%. In this range, the overall bias margins of a system containing an ERSFQ load and a FJTL primarily depend upon the actual bias current supplied by the bias network to the load. The average magnitude and variation of the different currents within a circuit are therefore extracted to observe behavioral trends and to more efficiently estimate the resulting bias margins.

Under ideal conditions, a FJTL is only used to establish a voltage reference. In nonideal conditions, a FJTL compensates the bias current in the load by redistributing (sacrificing) bias current to the load. To investigate these capabilities, an ERSFQ bias network is evaluated in both overbiased and underbiased conditions.

The underbiased condition corresponds to the case where the supplied bias current is lower than the target design objective. In the underbiased condition, reduced bias current in the FJTL does not initially affect the operation of the FJTL until the bias current is below the lower bias margin. At this current, the FJTL ceases to function properly (missed SFQ pulses or the absence of any switching in the FJTL). This underbiased FJTL produces a bias voltage that is lower than necessary. While several skipped pulses may not significantly change the bias voltage, a sufficiently underbiased FJTL terminates operation, making a reliable voltage reference no longer available. The underbiased load, depending upon the bias current, either exhibits increased delay or ceases operation.

The overbiased condition occurs when the supplied bias current exceeds the target design objective. In the overbiased condition, increased bias current in the FJTL does not affect the operation of the FJTL until the bias current is above the upper bias margin. At this current, the FJTL increases the switching rate, producing additional SFQ pulses. This overbiased FJTL produces a bias voltage that is higher than necessary. While this regime of operation does not affect the proper operation of the load, this system is highly energy inefficient. This inefficiency is due to the additional switching of the JJs in both the FJTL and the bias JJs within the load. The overbiased load, depending upon the bias current, exhibits either a slightly decreased delay or operates incorrectly.

A range of ERSFQ operation exists where the FJTL regulates the voltage and load current. While the circuit can operate outside of this range, the circuit either will exhibit suboptimal timing characteristics (the underbiased case) or is energy inefficient (the overbiased case).

This operational range is illustrated in Fig. 13.2 for a simple example—a FJTL connected to two JTL loads, one constantly switching due to the incoming data pulses and another not switching. Operation of a similar system had been modeled in [164]. In this figure, the dependence of the bias voltage on the supplied bias current is shown. The bias voltage is normalized to 25.85 μV ($\Phi_0 \times 12.5$ GHz)—a nominal bias voltage. The bias current is normalized to the target bias current of the system. The plateau around the bias current normalized to one (1) corresponds to the optimal range of operation of the ERSFQ-biased system. A higher bias current rapidly increases the bias voltage, expending significant energy. A lower bias current first produces another plateau, corresponding to the absence of switching in the FJTL. In this regime, the switching part of the load produces a lower bias voltage. Finally, if the bias current is further lowered, the load ceases to operate. In Sect. 13.3, similar under- and overbiased conditions are evaluated for different design parameters that would increase the range of ERSFQ operation.

Fig. 13.2 Regions of operation of ERSFQ system

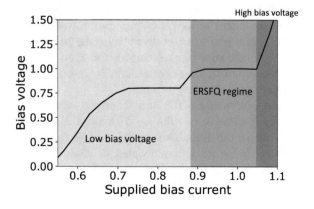

13.3   Trends and Guidelines of ERSFQ Bias Networks

In this section, certain parametric trends characterizing a bias network are discussed, and guidelines for ERSFQ bias network design tools are proposed. The system evaluated in this section is described in Sect. 13.2 and shown in Fig. 13.1. A metric frequently used in this section to evaluate these parametric trends is the dependence of the bias current supplied to the load on the bias current supplied to the system. The actual load bias current in this case is normalized to the design (target) load bias current—the sum of the critical currents of all of the bias JJs within the load. This normalization enables a comparison of the actual bias current to the bias margins of the load. The bias current supplied to the system is normalized to the sum of the target load bias current and the target FJTL bias current—the sum of the bias currents of all of the FJTL JJs when biased at 0.7 of the critical current. This approach enables the simulation of the under-/overbiased condition of the entire ERSFQ-biased system to be normalized. The primary focus of this section is to evaluate the behavior of the FJTL in terms of these parametric trends; slight variations in the extracted bias current are due to the redistribution of the bias current from the load to the FJTL. In Sect. 13.3.1, the effect of the bias inductance on current variations is compared to theoretical expectations. In Sect. 13.3.2, two different topologies of a FJTL stage are considered in terms of the bias distribution and energy efficiency. In Sect. 13.3.3, the effect of the bias margins of a FJTL on the overall system-wide bias margins is discussed. In Sect. 13.3.4, the effect of the size of the FJTL on the bias distribution network is evaluated, and design guidelines for the preferable size of the FJTL are suggested. In Sect. 13.3.5, the effect of the inductance of the source-to-FJTL path and source-to-load path on the supplied current in the presence of transient supply variations is described, and an approach to increase the inductance of the source-to-FJTL path is discussed.

13.3.1   Bias Inductance

ERSFQ gates are connected to a bias bus through large bias inductors. These inductors filter the high frequency current variations and reduce the amplitude of the bias current ripple, thereby reducing the probability of erroneously switching the JJs within the logic gates. A comparison of an analytic expression of the magnitude of the current ripple to simulations is described in this subsection to verify the correctness of the analysis process.

The dependence of the current variations on the bias inductance is illustrated in Fig. 13.3, where zero on the vertical axis is the average bias current. The overlapping plots depict the deviation from the average current for three different FJTL sizes. As noted in Fig. 13.3, the simulated variations in bias current are in good agreement with the theoretical value of Φ_0/L_B [156] and with the simulation results described in [163]. Similar to [163], any additional inductance more than 200 to 300 pH

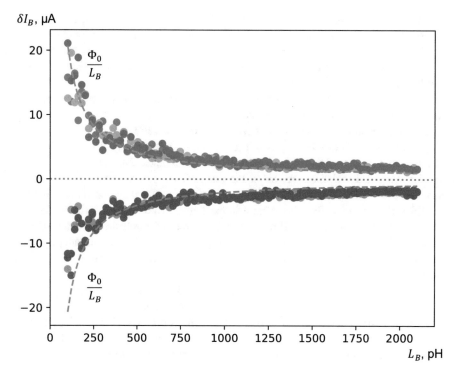

Fig. 13.3 Dependence of current variations on the bias inductance. The dashed line is the analytic expression

produces a negligible decrease in bias variations as compared to a typical bias current of an RSFQ gate. This example supports the application of this simulation analysis methodology to more complex parametric analyses.

13.3.2 Topology of FJTL Stage

One of the primary decisions in the automated synthesis of ERSFQ bias networks is the topology of the FJTL stage. Two methods exist for designing these structures—with [33, 163] and without [156, 367] a bias limiting JJ within the JTL stage. Different damping conditions for this bias JJ should be considered [163]. These parameters affect the size and efficiency of a FJTL and are therefore discussed in this section.

In underbiased circuits, no effect occurs from the presence of bias limiting JJs within a FJTL. As the bias of each individual stage is lower than the critical current of the bias limiting JJ, these JJs never switch, only slightly adding to the bias inductance as well as significantly increasing the area.

In overbiased circuits, the bias limiting JJs in both the FJTL and load continuously switch. This behavior increases the energy dissipation and decreases the
ability of the FJTL to absorb any excess bias currents. Without bias limiting JJs,
however, the bias current supplied to the FJTL can exceed the upper bias margin of
the FJTL, increasing the switching rate of the junctions within the FJTL, which in
turn dissipates more dynamic energy. A nontrivial tradeoff among the area, energy,
and bias regulation capability therefore exists.

A comparison of the bias regulation capability for a FJTL without a bias JJ, as
well as a FJTL with different sizes of the bias JJ, is shown in Fig. 13.4. The bias JJs
are assumed to either be critically damped or overdamped. Note that no difference
in bias current distribution occurs in the underbiased circuits. For the overbiased
circuits, the FJTL without bias JJs produces a preferable bias distribution for the
overbiased case (a smaller bias current in the load). From Fig. 13.4, the FJTL with
overdamped bias junctions follows the same trend, although the bias distribution
with overdamped bias JJs is improved as compared to a FJTL with critically damped
bias JJs. A FJTL without bias JJs in an overbiased circuit diverts more bias current
into the FJTL.

A comparison of the dynamic energy dissipation is shown in Fig. 13.5. The
dynamic energy is based on the total number of JJ switching events during a fixed
time period for different bias levels. FJTLs without bias JJs dissipate less energy
for all reasonable bias current levels (below 20% overbiasing), despite the higher
switching activity of the FJTL operating in the overbiased region.

13.3.3   Bias Margins of FJTL

The purpose of a FJTL in an ERSFQ circuit, apart from providing a voltage source,
is to absorb excess bias current in the overbiased circuits and to provide additional

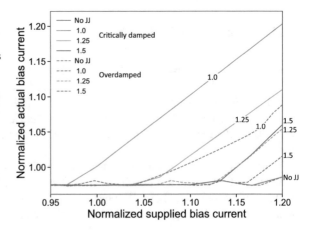

Fig. 13.4 Dependence of load bias current on supplied bias current for the critically damped and overdamped bias JJs within a FJTL. The example FJTL contains 64 stages, corresponding to 50% of the load bias current. The critical current of the bias JJ is normalized to 350 μA. Three different sizes of bias JJs (standard size, 25% greater size, and 50% greater size) are considered, as well as the absence of a bias JJ

Fig. 13.5 Dependence of total number of bias JJ switches and FJTL switches on supplied bias current. The critical current of the bias JJ is normalized to 350 μA. Three different sizes of bias JJs (standard size, 25% greater size, and 50% greater size) are considered, as well as the absence of a bias JJ

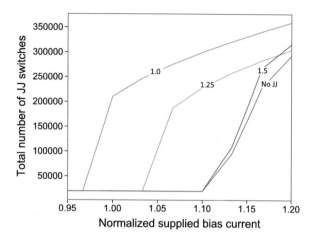

bias current to the underbiased circuits. A FJTL therefore experiences large current variations as part of the intended behavior. Despite a FJTL behaving as an analog voltage reference, correct operation of a FJTL depends upon the ability to pass SFQ pulses. Bias variations can produce either additional SFQ pulses or insufficient SFQ pulses to establish the correct bias voltage. A FJTL is therefore considered in this subsection as a digital transmission line, and the effects of the bias margins of the FJTL are discussed.

In overbiased circuits, a FJTL can switch more frequently, raising the voltage on the bias bus, expending additional energy. In underbiased circuits, a FJTL can either skip multiple SFQ pulses or completely cease operation, resulting in a loss of the voltage source and incorrect bias distribution. Wider FJTL bias margins improve the energy efficiency of the overbiased circuits. In the underbiased circuits, wider FJTL bias margins can increase the bias current in the load, ensuring the circuit operates properly at lower bias levels.

As confirmed in Fig. 13.6, wider bias margins of a FJTL improve the distribution of the bias current in the underbiased circuits and enable correct operation with a lower supplied bias current. The benefits of higher FJTL bias margins diminish beyond a bias margin of about 40%.

13.3.4 Size of FJTL

The size of a FJTL—the total bias current of all of the JTL stages comprising a FJTL—is another important parameter in ERSFQ bias networks. This current is typically compared to or normalized to the bias current of the load connected to the FJTL. With the JTL cells as part of a standard cell library, the total FJTL bias current is set by the number of stages. A larger FJTL can provide or absorb more current from the load and therefore better distribute the bias current. Additional

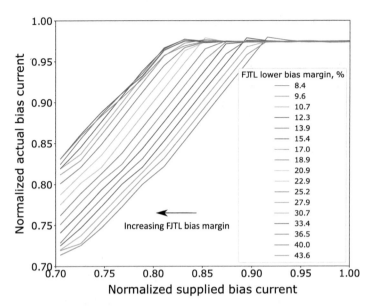

Fig. 13.6 Dependence of load bias current on supplied bias current for FJTL with the same size and different bias margins (8.4% to 43.6%)

stages, however, increase area and energy dissipation. In this subsection, those factors affecting the reasonable size of a FJTL for large scale circuits are discussed.

Combinations of different FJTLs and loads are evaluated for different supplied bias currents, as presented in Fig. 13.7. Note that a larger FJTL produces a preferable bias distribution in all cases, with the bias current closer to the target objective (smaller for overbiased cases, larger for underbiased cases). The benefits of increasing FJTL size, however, diminish with additional stages.

To constrain the size of a FJTL and provide effective design guidelines, two additional design parameters are necessary. One parameter is the target bias variations—the maximum bias current variations of an ERSFQ system including the FJTL. This target is constrained by those circuits with minimum bias margins within the load (least robust). An ERSFQ system is designed to guarantee to not exceed this bias margin.

The second design parameter is the maximum variation of the supplied bias current. This current is determined by the characteristics of the external bias source and the physical layout. In ERSFQ circuits with current recycling [51, 368], the maximum variation is the difference in bias current between serially biased circuit partitions.

The dependence of the size of a FJTL on the maximum variation of the supplied bias current and target bias current variations is shown in Fig. 13.8. The benefits of adding more stages to a FJTL diminish after the FJTL bias current exceeds approximately 50% of the load bias. In those cases where large bias variations are not expected, the bias current of a FJTL can be as small as 0% to 10% of the load bias

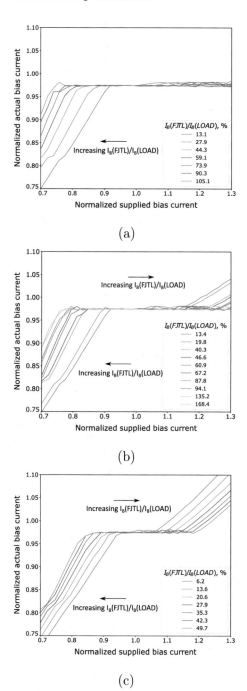

Fig. 13.7 Dependence of load bias current on the supplied bias current, (**a**) 8 bit load (21.31 mA), (**b**) 16 bit load (44.27 mA), and (**c**) 32 bit load (90.19 mA)

Fig. 13.8 Dependence of
FJTL size on desired bias
variations and expected
supply variations

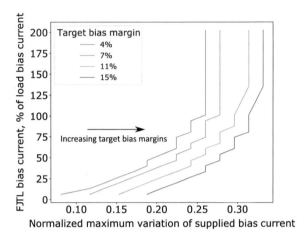

Normalized maximum variation of supplied bias current

current. For overbiased circuits, larger FJTLs further increase the range of energy-efficient operation, enhancing the robustness of the system against larger variations.

13.3.5   Inductance of Bias Bus

In ERSFQ bias networks, the bias bus is the line connecting a feeding JTL and load to a current source (the off-chip bias source connected to the on-chip I/O port). The inductance of this bias bus, typically parasitic, consists of two parts—the inductance L_{LOAD} between the current source (the I/O port) and the load and the inductance L_{FJTL} between the current source and the FJTL. This topology is schematically depicted in Fig. 13.1. These inductances are extracted from the physical dimensions of the bias lines. In this subsection, the effects of the inductance of this bias bus are discussed.

L_{LOAD} and L_{FJTL} form an inductive current divider, splitting the bias current between the FJTL and the load. While the steady-state current distribution depends upon the average voltage on the bias bus and the bias node within the cell, L_{LOAD} and L_{FJTL} affect the transient current distribution. By modifying the ratio of this current divider, additional bias current can be steered into the load in the case of a sudden decrease in bias current. A FJTL is among the most robust RSFQ circuits and exhibits wide parameter margins. Moreover, the output of a FJTL is typically not used, and therefore errors do not directly affect the circuit function. By redistributing additional bias current from the FJTL into the load, the function is preserved despite the system being temporarily underbiased.

Although the bias current eventually converges to the same current for all values of inductance, this approach improves the transient operation of the underbiased circuits in the presence of bias current variations and preserves the correct circuit operation when the bias current is temporarily reduced. This approach, however, further reduces the bias current supplied to a FJTL, partially sacrificing the high bias margins of the FJTL.

To isolate the load from changes in the supplied current, the inductance of the source-to-load path L_{LOAD} is increased. To enable a FJTL to accommodate changes in the bias current, the inductance of the source-to-FJTL path L_{FJTL} is decreased. These conditions can be converted into physical guidelines for bias network placement algorithms. The load should be placed farther from the I/O port, while the FJTL should be placed closer to the source. Alternatively, the wire shape can be modified to increase the inductance of the source-to-load path or an additional inductor can be added.

While the effect described here could be achieved by increasing the bias inductance of the load as compared to the bias inductance of the FJTL, an additional external inductance produces a less complex circuit layout. The bias inductors are connected in parallel. A large L_{LOAD} translates into a much higher L_B within the load, requiring more physical area.

Redistribution of bias current from a FJTL does not initially affect the operation of the FJTL until the bias current is below the bias margin, where the FJTL ceases to function properly (missed SFQ pulses or the absence of any switching in the FJTL). While several skipped pulses may not significantly change the bias voltage, a sufficiently underbiased FJTL terminates operation, removing the voltage reference from the system. The proposed approach is therefore constrained by the bias margins of the FJTL. An analysis of the proposed methodology is illustrated in Fig. 13.9. In Fig. 13.9a, the supplied bias current of the system momentarily changes to 90% of the target bias current. In Fig. 13.9b, the supplied bias current of the system momentarily changes to 110% of the target bias current. In Fig. 13.9a and b, respectively, the minimum and maximum bias current in the load are shown, normalized to the target load bias current.

The inductances, L_{FJTL} and L_{LOAD}, are varied from 5 pH to 50 pH. These inductances are chosen, somewhat arbitrarily, to demonstrate the proposed approach. The small inductance case is chosen as 5 pH, where the inductance is assumed to be mostly parasitic, and the number of bias inductors connected in parallel is large. The large inductance is limited to 50 pH to reduce simulation time. The size (bias current) of the FJTL is also varied from 13% to 50% of the load bias current. Currents closer to the target bias current are indicative of a more beneficial parameter set. The less shaded areas in Fig. 13.9 correspond to a higher bias current in the load—a desirable condition for underbiased circuits and an undesirable condition for overbiased circuits.

From Fig. 13.9, the difference between the highest and lowest bias current for different values of inductance in both underbiased and overbiased circuits can be over 11% of the load bias current for the case with a large FJTL (50% of the load bias current). For smaller FJTLs, this difference in the load bias current, dependent on the inductance, is smaller but over 6%. The choice of an optimal inductance is therefore important to improve the distribution of bias current in the presence of variations in the bias supply. As illustrated in Fig. 13.9, it is indeed desirable to increase the inductance of the source-to-load path and decrease the inductance of

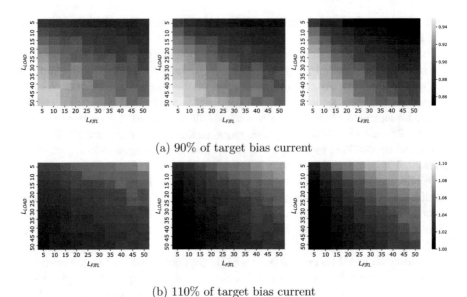

(a) 90% of target bias current

(b) 110% of target bias current

Fig. 13.9 Bias current in the load normalized to the target bias current in the load. The bias current supplied to the system is momentarily changed to (**a**) 90% and (**b**) 110% of the nominal current. The three different FJTL sizes (bias currents) correspond to, respectively left to right, 12.5%, 25%, and 50% of the load bias current. The brighter areas represent higher current

the source-to-FJTL path to improve the behavior of ERSFQ bias networks during transient changes in the supplied current.

13.4 Distributed FJTL Methodology

In this section, a methodology for distributed FJTL placement in large scale ERSFQ circuits is proposed based on the guidelines described in Sect. 13.3. Current design approaches commonly utilize one large FJTL for the entire IC [161]. This approach is suitable for small and medium scale ICs, enabling fast design of operational FJTLs and less complex testing [53, 54]. In addition, this approach is commonly used in test circuits, where additional control and signal inputs are available for the bias network.

A single FJTL, however, is not optimal for VLSI ERSFQ circuits. In these circuits, the desirable size of a FJTL can be on the order of hundreds of thousands of stages, requiring large on-chip area. In large scale circuits, different components operate at different clock frequencies [48]. A single common FJTL for an entire IC would necessarily operate at the highest of these frequencies, dissipating unnecessary dynamic power. As discussed in Sect. 13.3.4, the size of a FJTL depends upon the target bias margins. Different blocks and components within a complex circuit exhibit different target bias margins, resulting in different sizes for the corresponding FJTL. Requiring the strictest target bias margins from all FJTLs unnecessarily overdesigns the bias network, increasing both the physical area and dynamic power dissipation.

Furthermore, as suggested in Sect. 13.3.5, the inductance of the source-to-load path should be higher than the inductance of the source-to-FJTL path. In modern ERSFQ ICs, each input supplies only a small fraction of the total required on-chip bias current (a few hundred milliamperes). In prospective VLSI circuits, this fraction of the on-chip bias current will further decrease as the bias requirements increase. In circuits with a single FJTL, current from multiple bias pins is combined to provide the overall current. This shared bias bus spans the entire IC. Parasitic inductances within this bus are difficult to control and balance, making this topology inappropriate for large scale circuits.

In the proposed methodology, a VLSI ERSFQ circuit is subdivided into smaller blocks, consisting of circuits of similar complexity (and therefore similar target bias margins) and clock frequency. A similar approach is described in [164]. Each partition is connected to a separate FJTL and one or more independent bias pins. This approach enhances the control of both the FJTL characteristics and the parasitic inductances within each block, preventing overdesign of the bias network. Each distributed FJTL is connected to the clock source within the corresponding block, thereby operating at the correct local clock frequency.

Consider a large scale ERSFQ circuit with 20,000 gates divided into 2 different clock domains of equal size with a clock frequency of 20 and 50 GHz. Assuming each gate requires an average bias current of 500 μA, the total bias current is approximately 10 amperes. If the necessary bias current of the FJTL for this circuit is 15% of the total bias current (1.5 amperes) and the bias current of each FJTL stage is 350 μA, 4,285 FJTL stages (8,570 JJs) would be required. In the case of a single lumped FJTL, this topology would dissipate a dynamic power $P_D \approx (175\ \mu A) * \Phi_0 * (50\ \text{GHz}) * 8570 \approx 155\ \mu W$. In the case of two separate FJTLs, each operating at the corresponding clock frequency (20 and 50 GHz), $P_D \approx 77.5\ \mu W + 31\ \mu W \approx 109\ \mu W$, an approximate 30% improvement. This example illustrates the savings in dynamic power from the proposed methodology.

The disadvantage of the proposed methodology is the significant increase in circuit complexity. Multiple FJTLs operating at different frequencies require separate connections to corresponding bias networks. Separation of a large scale circuit into multiple clock domains requires separate clock and bias networks as well as synchronization elements for crossing clock domains [52].

Furthermore, this approach can be combined with current recycling techniques [369], where each partition is placed on a separate ground plane to reduce the total bias current. Partitions with equal bias currents are necessary for current recycling. FJTLs connected to a clock line can balance the bias requirements of different partitions while maintaining zero static power dissipation [51].

13.5 Conclusions

In this chapter, the bias distribution network for a cryogenic electronics technology—ERSFQ logic—is discussed. Robust bias networks are essential for the integration of ERSFQ circuits into LSI and VLSI complexity systems.

For different components within an ERSFQ bias network, trends are considered and advantageous tradeoffs are discussed. These design guidelines enable more robust ERSFQ circuits resistant to severe variations in bias current. Trends and tradeoffs described in this chapter provide a means to decrease the bias current of FJTLs within large scale ERSFQ circuits, thereby reducing physical area and power dissipation.

A methodology for the distributed placement of ERSFQ FJTLs in large scale circuits is described. This methodology enables precise control of the parasitic inductances within a bias network, reducing area and power as compared to a single large FJTL. The proposed methodology and guidelines can be integrated into commercial EDA bias network design tools for prospective ERSFQ VLSI circuits, incentivizing SFQ as a promising beyond CMOS technology.

Chapter 14
Partitioning RSFQ Circuits for Current Recycling

RSFQ circuits require a DC bias current to operate properly. The bias current in conventional RSFQ circuits is supplied to each gate, resulting in large current requirements in VLSI complexity SFQ systems, on the order of tens to hundreds of amperes. These high currents are difficult to supply and distribute. Superconductive input/output pins and bias lines support this limited current. Large currents however require significant metal and input pin resources. In addition, large currents can inductively couple to sensitive superconductive inductors, degrading circuit operation and producing errors. Current recycling is a well known technique to reduce these bias currents. RSFQ circuits with similar bias current requirements can be placed on separate ground planes and serially biased. The inputs and outputs of these circuits are galvanically decoupled and require drivers and receivers between connections.

In this chapter, a methodology for automated partitioning of complex RSFQ circuits into blocks with similar bias currents is described, where the number of connections among the blocks is minimized. These blocks are biased in series, reducing the total bias current by the number of partitions. The partitioning methodology is intended for use within an automated EDA flow to enable current recycling for arbitrary (nonuniform, irregular) VLSI complexity RSFQ circuits, drastically reducing overall bias current and input requirements [51].

The chapter is organized as follows. In Sect. 14.1, a motivation for reducing the total bias current in superconductive circuits is provided. In Sect. 14.2, current recycling (serial biasing) in RSFQ circuits is briefly introduced, and existing challenges are described. In Sect. 14.3, a methodology for partitioning RSFQ circuits into blocks with similar bias current requirements during the placement process is presented. This chapter is concluded in Sect. 14.4.

© Springer Nature Switzerland AG 2022
G. Krylov, E. G. Friedman, *Single Flux Quantum Integrated Circuit Design*,
https://doi.org/10.1007/978-3-030-76885-0_14

14.1   Introduction

Existing and prospective fabrication technologies enable VLSI complexity SFQ circuits with a complexity of over a million JJs per die, approximately several hundred thousand gates [112]. Assuming each gate requires a bias current of several hundred microamperes, the total estimated bias current is on the order of tens to hundreds of amperes. These high currents are difficult to efficiently supply and distribute [78]. Supplied from off-chip, these currents also require a large number of input pins. As each input has a current limit of 200–300 mA, the total number of inputs required to supply the bias system can exceed hundreds. In addition, the bias current is distributed on-chip though thin superconductive lines, which are necessarily wide to support these high currents, expending limited metal resources. Furthermore, the high currents produce large magnetic fields which can couple to sensitive RSFQ gates, introducing errors [54]. It is therefore imperative to reduce on-chip bias currents within complex RSFQ circuits [48].

14.2   Current Recycling

A well known technique to reduce the total bias current is current recycling, also known as serial biasing [368, 369]. In this technique, the circuit is partitioned into multiple segments (or islands) with approximately the same bias current. These segments are galvanically isolated from each other and the rest of the circuit. The partitions use different ground planes and are serially biased, with the ground of each partition acting as a bias supply for the next partition. The inputs and outputs of these serially biased islands are connected to the rest of the system through special pulse transfer circuits, requiring inductive or capacitive coupling [370]. These couplers (driver-receiver pairs) typically consist of a few JJs and coupling elements and occupy significant area [371]. It is therefore desirable to reduce the number of these couplers.

Serial biasing was first proposed in 1989 [372] and is widely used in complex RSFQ circuits [373]. This approach however requires manual ad hoc designation of repetitive blocks within a circuit—a process difficult to automate and reuse. Unlike highly regular, specialized circuits typically used to demonstrate current recycling (e.g., shift registers), a limited number of repetitive structures exists in general VLSI circuits. The benefits of current recycling in this case are applicable to only a small portion of a system. Moreover, automated design tools for RSFQ circuits, in particular, automated place and route (APAR) tools, need to be aware of these current recycling features, as the placement and routing process changes with the introduction of separate ground planes and driver-receiver pairs [107].

To date, current recycling has not been applied to general RSFQ circuits. In this chapter, a methodology is proposed to mitigate these issues by automatically partitioning arbitrary logic to support current recycling [374]. A similar methodology has been simultaneously and independently proposed in [375], where the ground plane is partitioned after the circuits have been placed (post-placement). The primary distinctive feature of this work is that partitioning is performed during the first steps of the coarse placement process, when the location of the gates has yet to be finalized. In addition, different partitioning and optimization algorithms are used.

14.3 Partitioning of Arbitrary RSFQ Circuits During Placement

In this section, a methodology for partitioning RSFQ circuits during placement is proposed and demonstrated. In Sect. 14.3.1, serial biasing of partitions with slightly different bias currents is discussed. In Sect. 14.3.2, partitioning during placement is introduced and the advantages and disadvantages of this approach are reviewed. In Sect. 14.3.3, the quadratic placement algorithm used here for coarse placement is briefly introduced. In Sect. 14.3.4, partitioning RSFQ circuits with the Fiduccia-Mattheyses (FM) algorithm is discussed, and results are presented for a number of benchmark circuits. In Sect. 14.3.5, geometric partitioning with simulated annealing is described, and results are presented and compared to partitioning with the FM algorithm.

14.3.1 Unbalanced Partitioning of RSFQ Circuits with Padding

In existing RSFQ circuits utilizing serial biasing, all of the partitions exhibit identical bias currents [373]. This condition, however, is not a strict requirement. While having different bias currents for each island degrades the bias margins, resulting in over- or underbiasing of the entire island, certain small differences in the bias current can be tolerated [154]. The balance conditions for the partitioning algorithm therefore depend upon the robustness of the circuit to changes in the bias current.

RSFQ technology provides a means to mitigate any imbalance among the different islands. Dummy gates, for example, a chain of JTL stages or individual JJs, can be added to those islands with a smaller bias to equalize the bias current among islands. A padding JTL can be placed anywhere within an island and only needs to be connected to a bias line within this island, not a signal line. This JTL does not need to be operational or efficient. The total bias current of these elements can therefore be arbitrarily chosen to equalize the bias currents among the islands while reducing the added area.

In ERSFQ circuits [33], it is desirable to connect a dummy JTL to a clock line. In this case, due to the properties of ERSFQ bias networks, as discussed in Chap. 13 [166], padding elements do not dissipate static power. Furthermore, ERSFQ circuits utilize a feeding JTL (FJTL) as a voltage regulator [49]. In an ERSFQ circuit with current recycling, the FJTL sets the bias voltage for each island. Each island therefore requires a separate FJTL [163]. Alternatively, a single FJTL can be located within the island with the highest ground voltage. In this way, the FJTL provides a reference for the highest voltage in the circuit although dissipating additional dynamic power. The current regulation capability of a FJTL also provides additional robustness to small variations in bias current. If the circuit is slightly over- or underbiased, the FJTL compensates for these small changes, maintaining the bias current in the corresponding circuit close to the design target [50].

14.3.2 Partitioning During Placement

Any complex circuit can be represented as a hypergraph, with vertices corresponding to the gates and hyperedges corresponding to the connections between the gates. Hypergraph partitioning algorithms and heuristics are both widely used in CMOS EDA placement tools [376]. These algorithms are included within the coarse placement process to minimize the number of connections to the different parts of a circuit, thereby reducing the total wire length and overall wiring congestion. Balanced partitioning of a hypergraph is an NP-hard problem [376]. A variety of heuristics have therefore been developed to enable the partitioning and placement of VLSI circuits.

The primary advantage of partitioning during placement is integration into existing EDA flows. The placement tools perform partitioning and can be modified to consider the bias current of the cells during the placement process. In addition, empty space can be reserved by the placement tools to place and route driver-receiver pairs between islands. The number and area of these structures depend upon the number of connections between different partitions, and are not known in advance. This approach also avoids complex irregular shapes for the ground planes, which can be difficult to characterize. With the additional constraints of balancing the bias currents among partitions, partitioning during placement can increase the average wire length [377].

Unlike partitioning during placement of large scale CMOS circuits, a small number of partitions for current recycling in RSFQ provides significant benefits. The total bias current of a circuit separated into N partitions is reduced by N. The number of connections between islands also increases with the number of islands. The total number of partitions for current recycling can therefore be small (up to a few tens of partitions). An advantage of recursive bipartitioning is that each pair of partitions is balanced by the bias current, enabling reuse of the same padding elements.

Multiple modifications are necessary to apply existing partitioning algorithms to the problem of RSFQ current recycling. The reduced number of connections between partitions produces fewer driver-receiver pairs. Moreover, due to the limited fanout, RSFQ circuits can be represented as a regular graph rather than as a hypergraph, where each net (edge) is a one-to-one connection between two gates. In CMOS, the balance optimization condition is typically related to the area of the gates [378]. For current recycling in RSFQ circuits, the bias currents are the primary issue within the balance condition, as each gate requires a specific bias current.

14.3.3   Coarse Placement

Many algorithms exist for the automated placement of standard cells [240]. The partitioning methodology presented here is embedded into the first stages of the coarse placement process. During this stage, an approximate location is determined for each cell based on wire length constraints. These locations often significantly overlap. This issue is later resolved in the placement process during the legalization step.

The quadratic placement (QP) algorithm is used here to produce a coarse placement for the proposed methodology. QP is a well known and relatively old algorithm for analytic placement [379]. Among the primary advantages of QP are low computational complexity and global minimization of the wire length. Although many cells are frequently placed at the same coordinates, this placement procedure produces a high fidelity representation of the approximate gate locations. In this section, the QP algorithm is briefly described.

The primary objective of quadratic placement is to minimize the wire length. A connectivity matrix $C = [c_{ij}]$ is produced, where $c_{ii} = 0$ and c_{ij} represents the weight of the connection between different nodes (gates or standard cells) [380]. Different models exist for the nets connecting multiple nodes [379]; for RSFQ logic, all nets can be treated as one-to-one nets. The cost function to minimize the wire length is

$$F = \frac{1}{2} \sum_{i,j=1}^{n} c_{ij}((x_i - x_j)^2 + (y_i - y_j)^2). \tag{14.1}$$

An auxiliary diagonal matrix D is introduced, where $d_{ii} = \sum_{j=1}^{n} c_{ij}$ is the sum of all of the connection weights for a target node [380]. The objective function (14.1) is rewritten in the form,

$$F = x^T Q x + y^T Q y, \tag{14.2}$$

where $Q = D - C$ and x and y are the coordinates for each gate. This problem is reformulated as two linear systems in x and y coordinates (only x is shown here for brevity) [380],

$$Q_{cc}x_c = -Q_{cf}x_f, \tag{14.3}$$

where Q_{cc} and Q_{cf} are submatrices of Q corresponding to the connections, respectively, between the nodes and between the nodes and immovable pads. x_c and x_f are, respectively, the coordinate of the nodes and pads [379]. The coordinates can be determined using any method for solving a system of linear equations. As each node is connected to only a few other nodes, Q is sparse, resulting in high computational efficiency of QP even for a large number of nodes. The QP algorithm is used here to produce a high fidelity model of the initial coarse placement.

14.3.4   Partitioning Using Fiduccia-Mattheyses Heuristic

In this section, the Fiduccia-Mattheyses algorithm [308], an improvement on the earlier Kernighan-Lin algorithm [381], is used for bipartitioning. The number of connections between partitions is minimized, while the balance conditions for the partitions are modified to consider the bias currents of different nodes (gates/cells).

For partitioning using FM, a few important terms need to be introduced. A cut is an edge connecting vertices in different partitions. In RSFQ circuits with current recycling, each cut corresponds to a connection between different serially biased blocks. Each cut therefore requires a coupler (driver-receiver pair) circuit. For each vertex, a gain corresponds to a change in the number of connections, resulting from the movement of this vertex into a different partition. If all neighboring vertices of a given vertex are already in the updated partition, the gain is maximum. The vertex with the highest gain represents the best possible move.

The FM algorithm consists of multiple passes. Each pass consists of multiple iterations (moves) and operates as follows. The graph is initially separated into two random partitions. The gain is calculated for each vertex. The move with the highest gain is selected. The corresponding vertex is moved to the other partition. The gain values for the neighboring nodes are updated, and the vertex is locked—it cannot be moved again during this pass. This procedure is repeated until all of the vertices are locked. The best partition during a specific pass is used as the initial partition for the following pass, and all nodes are once again unlocked. The algorithm terminates when no further improvement in the solution is produced. The balance conditions are introduced to avoid moving all gates into one partition. The vertex is only moved if the resulting partitions are balanced.

RSFQ technology can also provide certain benefits to the FM algorithm. Most RSFQ gates exhibit a fanout of one, and most cell libraries contain only two input logic gates. Including the clock signal, the highest/lowest gain is bounded by ± 4 (number of inputs and outputs). Unlike CMOS, no nodes exist with a larger gain including the clock input.

Partitioning during the placement process is illustrated in Fig. 14.1. The initial circuit netlist, in graph representation, is placed using the QP algorithm, as shown in Fig. 14.1a. The circuit is bipartitioned using the FM algorithm. When partitioning

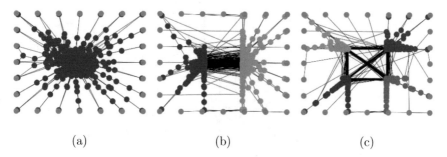

Fig. 14.1 Coarse placement and partitioning with FM heuristic of a benchmark circuit, (**a**) initial placement, (**b**) bipartitioning and placement, and (**c**) four-way partitioning and placement. Different shaded circles (nodes) correspond to individual cells in different partitions, while the lines (edges) represent signal connections

RSFQ netlists for current recycling, the primary balance condition is the bias current. In addition, the gain of the node can be modified to consider the relative imbalance among partitions and the approximate coordinates determined by the coarse placement algorithm. Coarse placement is performed on each of the resulting partitions, where the connections to the nodes in another partition are represented by immovable pads, as shown in Fig. 14.1b. As the QP algorithm aims to minimize wire length, the connected nodes within different islands are visibly "pulled" toward each other. The space between the partitions in the layouts, shown in Fig. 14.1b and c, is exaggerated. As discussed in Sect. 14.3.2, this spacing can be adjusted based on the number of driver-receiver pairs. This procedure is recursively repeated on subsequent partitions, as shown in Fig. 14.1c, until the target number of partitions is reached. In a standard cell design flow, the nodes are placed within the corresponding cell rows at later stages of the placement process.

This partitioning methodology is implemented in Python and has been evaluated on CMOS industry standard ISCAS'89 benchmark circuits [382] and the AMD2901—a 4 bit slice ALU. The benchmark circuits have been modified to better consider RSFQ circuits. The netlist characteristics for these circuits—the bias current and the number of nodes (gates) and edges (nets)—are listed in Table 14.1. Splitters are included for multiple fanout [151]. Multiple fanin gates are replaced by gates with only two inputs. NAND/NOR gates are divided into AND/OR and NOT gates, and the clock inputs are included in each logic gate. A clock distribution network composed of a large splitter tree is also introduced [169], as clock splitters comprise a significant fraction of the total bias current [52]. Two different AMD2901 ALUs are used—the AMD2901f which includes the necessary path balancing D flip flops and the AMD2901s which does not include these extra flip flops.

The results of applying recursive bipartitioning and placement to these benchmark circuits using this methodology are listed in Table 14.1. Note that the FM algorithm utilizes an initial random or semi-random partitioning and is therefore nondeterministic—each run produces slightly different results. This methodology

Table 14.1 Results of FM partitioning on modified ISCAS'89 benchmark circuits and the AMD2901 ALU

Benchmark circuit	Bias current, mA	Vertices (gates)	Edges (nets)	Two partitions				Four partitions			
				Cuts	Bias imbalance	Bias difference, mA	Maximum bias current, mA	Cuts (total)	Bias imbalance (maximum)	Bias difference (maximum), mA	Maximum bias current, mA
s27	27	49	70	8	4.8%	0.6	13.7	15	5.7%	0.4	6.9
s298	382	626	942	70	1.3%	2.5	192	125	1.4%	1.3	96
s344	379	620	950	75	1.5%	2.8	191	125	1.7%	1.6	96
s420	552	889	1354	96	1.1%	2.9	278	180	1.3%	1.8	139
s1238	1516	2450	3776	390	0.02%	0.2	758	657	0.1%	0.6	379
AMD2901s	2218	3021	4799	459	0.01%	0.3	1109	650	1.2%	6.7	558
AMD2901f	3313	9349	11,127	205	1.8%	30	1671	669	3.4%	27	842

minimizes the number of connections among islands while satisfying the required balance constraints. These results emphasize balancing the bias current among the islands. The number of connections among the islands can be further reduced if this objective is prioritized within the partitioning algorithm. As listed in Table 14.1, this methodology produces a reasonable number of connections between partitions for complex, highly irregular circuits. For two-way partitioning, the fraction of cut nets is approximately 5–15%, depending upon the structure of a particular circuit. With a larger number of partitions, the number of cut nets increases. Approximately 20% of all connections are cut for four-way partitioning. The resulting partitions can be serially biased, reducing the total bias current required by the system. For small differences in bias current, as listed in Table 14.1, dummy padding structures for bias balancing may not be necessary, particularly in ERSFQ circuits.

Although this FM heuristic produces highly balanced partitions with a small number of cut nets, it is computationally hard. The time required to partition a large circuit using the FM algorithm can become impractical. Clustering steps are typically added before partitioning to reduce the complexity [376].

14.3.5 Geometric Partitioning with Simulated Annealing

A major drawback of this partitioning process using the FM algorithm is the introduction of multiple steps into the design flow. Although APAR tools include multiple partitioning steps as part of the placement process, integration of additional constraints related to equalizing the bias current can increase the total wire length and runtime of the placement process. A simpler technique, less intrusive within established EDA flows, is geometric partitioning [383]. In this partitioning approach, coarse placement of a circuit is divided into multiple blocks based only on the cell (node) coordinates, as shown in Fig. 14.2a.

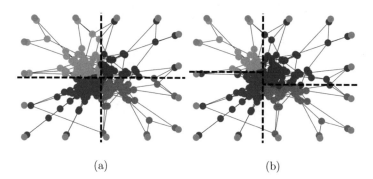

(a)   (b)

Fig. 14.2 Coarse placement and geometric partitioning of a benchmark circuit, (a) simple geometric partitioning with equal area, and (b) geometric partitioning with partition boundaries optimized by simulated annealing

Table 14.2 Results of geometric partitioning on modified ISCAS'89 benchmark circuits and the AMD2901 ALU

Benchmark circuit	Two partitions		Two partitions with optimization				Four partitions with optimization			
	Cuts	Bias imbalance	Cuts	Bias imbalance	Bias difference, mA	Maximum bias current, mA	Cuts (total)	Bias imbalance (maximum) mA	Bias difference (maximum), mA	Maximum bias current, mA
s27	6	158%	16	0.2%	0.03	13.4	25	14%	0.9	7.2
s298	27	1198%	104	0.4%	0.8	191	185	57%	42	114
s344	90	11%	90	0.4%	0.7	190	185	13%	11	100
s420	48	786%	117	1.4%	3.8	278	243	37%	46	170
s1238	194	739%	386	1.5%	12	764	738	25%	80	405
AMD2901s	177	109%	318	1.0%	11	1114	667	5.1%	27	560
AMD2901f	77	2365%	273	0.5%	8.1	1660	692	11%	90	881

As described in Sect. 14.3.3, the initial placement process aims to minimize the total wire length. Dividing this placement into partitions of equal area based only on the coordinates can produce severely unbalanced partitions with a large number of cut nets, as listed in Table 14.2 for the same benchmark circuits (as described in Sect. 14.3.4). If the slightly modified coordinates of the partition boundaries consider the bias currents of the resulting islands, as shown in Fig. 14.2b, the balance characteristics are improved.

Simulated annealing is used here to select the preferable boundary for the geometric partition, producing a smaller bias imbalance and fewer cut nets. For bipartitioning, only one boundary is considered, while for four-way partitioning, three boundaries are considered (one for the initial bipartition and one for each resulting partition). The results of geometric partitioning with simulated annealing are also listed in Table 14.2. As with FM partitioning, described in Sect. 14.3.4, the cost function for simulated annealing emphasizes a balance of the bias currents between different islands. Fewer cut nets with a higher bias imbalance can be produced by changing the cost function of the optimization process.

As seen from a comparison between Tables 14.1 and 14.2, geometric partitioning generally produces a greater bias imbalance between islands with more cut nets as compared to FM partitioning. This approach, however, is computationally more efficient and is less difficult to integrate into an established design flow. This approach also greatly reduces the maximum bias current of a system due to serial biasing while requiring minimal changes to the circuit at the cost of increased area.

14.4 Conclusions

A methodology for current recycling in large scale RSFQ circuits is described in this chapter. The methodology enables current recycling in complex irregular RSFQ circuits by partitioning these circuits into islands with a similar total bias current. The partitioning step is integrated into the first stages of the coarse placement process. Islands with a significantly different bias current are balanced using dummy padding structures. These structures may not be necessary for highly balanced partitions. A modified Fiduccia-Mattheyses algorithm is used for balanced partitioning to reduce the number of galvanically isolated drivers/receivers in an RSFQ circuit with serial biasing. Approximately 5–15% for bipartitioning and approximately 20% for four-way partitioning of the nets within a circuit require a driver-receiver pair, while the differences in bias current are on the order of a few percent. An alternative partitioning technique, geometric partitioning with optimization, is also described. This approach produces more unbalanced partitions, requiring additional driver-receiver pairs. It is however computationally more efficient and introduces fewer modifications into the placement process while enjoying the benefits of current recycling. This partitioning methodology can be integrated into standard CMOS-like EDA design flows for RSFQ circuits by including bias weighted partitioning during the placement process.

Chapter 15
GALS Clocking and Shared Interconnect for Large Scale SFQ Systems

A globally asynchronous, locally synchronous clocking scheme for large scale single flux quantum systems is described in this chapter. In this scheme, the width of each data bus is extended to carry the corresponding clock signal. This signal activates the distribution of the clock signals within the receiving block. Based on this approach for intra-chip interconnect within SFQ systems, a configurable shared bus is also proposed. The data are attached to a tag, and a resulting data packet is sent to the shared bus. This packet is received by each block but only processed if the tag matches the block identifier. By avoiding expensive comparators and multiplexers, the overhead of the global bus connection is reduced. The proposed approaches exploit the pulse-based nature and ambiguity of clock and data in SFQ technology—the data packet propagating through the interconnect carries a local clock signal [52].

15.1 Introduction

Recent advances and ongoing research efforts in the area of superconductive electronics, including the development of EDA tools and methodologies as previously discussed in Chap. 7, will greatly enhance the development of large scale single flux quantum systems [269, 366]. With improved integration and complexity of these superconductive systems, the number of functional units will significantly increase.

In modern CMOS systems, communication among the functional blocks within a system-on-chip is achieved using configurable interconnects [384]. These interconnects provide sophisticated interfaces for scheduling and ordering the requests for the different devices and memory. This approach supports modular design and reuse of IP cores with minimal modifications while significantly reducing overall development time. An additional benefit of this approach is the flexibility and

© Springer Nature Switzerland AG 2022
G. Krylov, E. G. Friedman, *Single Flux Quantum Integrated Circuit Design*,
https://doi.org/10.1007/978-3-030-76885-0_15

scalability of the resulting system—functional blocks can be added or removed without extensive modifications to the existing system [51].

Another important issue in large scale SFQ systems is timing. SFQ circuits are clocked at tens [133, 193] and even hundreds [34] of gigahertz, resulting in clock periods on the order of several to tens of picoseconds, as previously discussed in Chap. 5. Timing tolerances in these circuits are therefore extremely narrow, on the order of a few picoseconds. This issue will become more severe in future SFQ VLSI circuits, as a primary application is high speed computation. One commonly used technique to mitigate timing variations across functional blocks and to simplify global timing in CMOS is globally asynchronous, locally synchronous (GALS) clocking [170], as previously described in Sect. 5.1.3. In this approach, each functional block is locally clocked; a global clock distribution network is not required. Consequently, the area, power, and timing issues of global clock networks are avoided.

In this chapter, two novel approaches exploiting a pulse-based signal representation of both data and clock in SFQ circuits are proposed. The GALS clocking scheme provides a means for data packets to carry a local clock, avoiding the overhead of a global clock network. This topic is discussed in Sect. 15.2. The proposed bus is an extension of the GALS scheme and provides another means to exploit the ambiguity of clock and data in SFQ circuits to enhance asynchronous and self-timed communication while requiring reasonable overhead. This topic is described in Sect. 15.3. In Sect. 15.4, some simulation results are provided, followed by some concluding remarks in Sect. 15.5.

15.2   GALS Clocking Scheme for SFQ Circuits

In this section, specific features of SFQ are discussed, and means to exploit these features are proposed. In Sect. 15.2.1, the ambiguity of clock and data in SFQ is introduced, and early work exploiting this property is discussed. In Sect. 15.2.2, SFQ specific advantages for clock distribution are highlighted. In Sect. 15.2.3, a GALS clock activation approach utilizing this property is presented.

15.2.1   Ambiguity of Clock and Data

A distinctive quality of SFQ technology is the pulse-based nature of both data and clock, as previously discussed in Sect. 3.3. Information is encoded by the presence (logic "one") or absence (logic "zero") of an input SFQ pulse within a specific clock period. The logic gates process the inputs received during the clock period at the time of arrival of the clock pulse. The clock pulse triggers the junctions to produce an output, based on the inputs [48]. The clock pulses arriving at the clock input are indistinguishable from the data pulses arriving at the data inputs. The clock and data inputs within an SFQ logic gate can therefore be arbitrarily exchanged to achieve a desired behavior. Moreover, the clock signal can be locally regenerated from the data at each gate.

This distinctive feature has been explored in data-driven self-timed (DDST) RSFQ circuits [201]—a class of dual-rail SFQ circuits previously described in Sect. 5.2.2. In DDST circuits, the data are carried by complementary signals using two parallel lines for each bit. When data arrive at the next logic gate, the clock is generated by a logical OR function. At the output of the logic gate, the complementary signal is provided by a D flip flop with complementary outputs. While DDST circuits require significant overhead for routing as well as additional circuitry for generating complementary signals by each logic gate, this approach enhances controllability of the clock timing, improves robustness to process variations, and reduces the overhead of the global clock network.

Other important methods of a pulse-based data representation for clocking large circuits are the concurrent and counterflow clocking schemes, previously described in Sect. 5.1. In the concurrent scheme, the clock pulse travels together with the data through the logic pipeline. In the counterflow scheme, the clock pulse travels in the opposite direction relative to data. In the clock-follow-data scheme, which is a type of concurrent scheme, the clock signal follows the data, marking the end of a clock period.

15.2.2   Clock Generation and Distribution

Efficient on-chip clock generation is necessary to operate multi-gigahertz systems within a cryogenic environment, as each high speed connection to/from room temperature is costly. SFQ technology, as opposed to CMOS, provides multiple efficient ways for on-chip clock generation with low overhead. Every Josephson junction with an applied DC voltage generates a train of SFQ pulse through the Josephson effect [385]. Other methods for clock generation, such as arrays of junctions and long Josephson junctions, have also been proposed [385, 386]. For analog applications, the high Q factor of an on-chip oscillator is critical to reduce clock jitter. For digital applications, a simple ring oscillator is typically sufficient to generate a repetitive clock waveform.

Pulse-based signal representation leads to another beneficial property of SFQ circuits for clock generation and distribution. A switching Josephson junction regenerates the incoming SFQ pulse, restoring the shape of the signal. For a clock network within a functional block, the entire block can be clocked by a single clock pulse through a combination of ring oscillators and splitter gates. This property lessens the need for a global clock distribution network.

15.2.3   Clock Activation Scheme

A combination of DDST and clock-follow-data techniques—a GALS clock activation scheme—is presented here. In this scheme, each multi-bit signal connection between functional blocks is extended by one additional signal, carrying the clock. Within each block, the clocking scheme is individually chosen to satisfy local design criteria and can be one of several synchronous schemes [180, 181] (binary tree, concurrent, counterflow) or an asynchronous scheme, where the incoming clock signal serves as a handshake signal. The proposed approach exploits the pulse-based nature and ambiguity of clock and data in SFQ technology—the data propagating through the interconnect carry a local clock signal. This approach, depicted in Fig. 15.1, can be applied to any infrequently used circuit or block within a larger circuit that can benefit from the temporary absence of a clock signal.

The overhead of this approach is one additional signal line for every data bus. The area overhead of routing an additional line within the data bus is 3.1% for a 32 bit bus and 1.6% for a 64 bit bus.

The clock distribution network within the receiving circuit introduces a delay for each gate between the reception of the activation signal and generation of the appropriate clock signals. In addition, the driving circuit provides a clock activation signal for the receiving circuit that signals the end of the current operation. The overhead of the generating and activating circuits depends upon the particular circuit. The receiving circuit remains unchanged.

The benefits of the proposed GALS topology include a reduction in dynamic power dissipation and lower overall clocking complexity. In addition, this approach enables safe clock domain crossings [387] for SFQ systems with multiple clock domains.

Fig. 15.1 Proposed GALS scheme

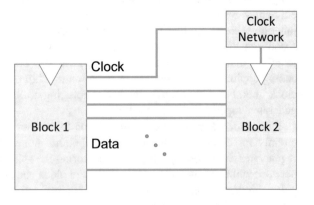

15.3 Shared Interconnect

Heretofore, the low complexity of SFQ circuits did not justify the significant overhead of advanced bus architectures and interconnect networks widely used in modern CMOS systems on chip. These interconnects are therefore frequently designed manually on an ad hoc basis. With higher complexity superconductive systems, the development of a structured interconnect network for SFQ circuits will become necessary. SFQ technology provides higher speed and lower power interconnects as compared to CMOS technology. In the following subsections, these advantages are described, and relevant applications for the proposed bus architecture are discussed. In Sect. 15.3.1, the primary types of interconnect for SFQ technology are briefly reviewed. In Sect. 15.3.2, the input discrimination characteristics of SFQ circuits, particularly beneficial for bus structures, are discussed. Based on these properties, in Sect. 15.3.3, a novel shared bus topology for SFQ circuits is proposed.

15.3.1 Types of SFQ Interconnect

Two types of signal interconnect used in SFQ integrated circuits are passive transmission lines (PTLs) and Josephson transmission lines (JTLs) [23, 107], as previously discussed in Sect. 4.1. A JTL is an active structure composed of biased and grounded JJs connected by inductive lines. SFQ pulses are regenerated at each stage of a JTL, expending energy and adding delay. A PTL is a passive stripline or a microstripline connecting a driver and a receiver. SFQ pulses in a PTL propagate ballistically over significant distances at the speed of light within the medium with negligible loss. These distances can exceed several millimeters before a repeater is needed [141, 146, 388]. Multiple pulses can also simultaneously propagate along a PTL (in a wave pipelined fashion, see Sect. 5.2.1), further increasing throughput.

15.3.2 Input Discrimination

Another important consequence of a pulse-based data representation of SFQ circuits is improved discrimination of unwanted inputs. Each input of an SFQ circuit supports multiple fan-in—multiple input lines from different sources which can be electrically connected to a circuit input, assuming no pulses simultaneously arrive. If the input registers support the rejection of multiple data pulses within a clock period through an escape junction or if blocking gates are used [53, 54], the erroneous data are constrained to just one clock period. This property is useful for bus structures, similar to three state buses in conventional CMOS circuits [389].

15.3.3 Bus Topology

In this subsection, a shared bus topology is proposed, exploiting the aforementioned properties of SFQ circuits and the GALS clock activation scheme described in Sect. 15.2. In this proposed bus topology, as depicted in Fig. 15.2, multi-bit data originating from each device are attached to a tag identifying the recipient block, forming a data packet, e.g., the memory access request is attached to a tag identifying the memory controller. The resulting packet is passed to the shared bus, consisting of passive transmission lines, Josephson transmission lines, and functional block interfaces containing a hard wired block identifier.

At each block interface, the tag in the incoming data packet is compared to the block identifier. As the identifier is predetermined, the comparator does not require significant overhead. For a matching condition, a control pulse is generated. This pulse is a write enable signal for the input register of the block, a handshake signal for an asynchronous block, or a clock signal for an entire block [47]. In the latter case, the data inputs are converted into clock signals, as described in Sect. 15.2.2—a feature not possible in CMOS.

The data arrive at all of the functional blocks connected to the bus and are stored within the input registers. Processing, however, is only initiated if the tag matches the identifier. In this way, fewer expensive decoders at each block interface are needed. To reach the previous block connected to this interconnect, the bus is a

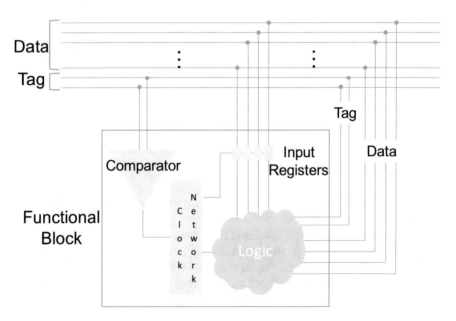

Fig. 15.2 Proposed shared bus topology

circular topology, with the end connected to the beginning. A similar approach has been previously proposed in a CMOS-based optical ring bus [390].

15.4 Behavioral Characteristics

In this section, simulation results describing the GALS clock activation scheme are presented. In Sects. 15.4.1 and 15.4.2, globally asynchronous communication between two synchronous shift registers is demonstrated, and certain timing characteristics are illustrated. In Sect. 15.4.3, several approaches for a clock activation methodology are discussed. In Sects. 15.4.4 and 15.4.5, respectively, the applicability of the proposed approach to multi-chip modules and energy efficient SFQ is reviewed.

15.4.1 Behavior of H-Tree Clock Networks

Two 64-bit-wide shift registers with a depth of 8 bits are generated via a Python script. These SFQ shift registers utilize an H-tree clocking topology [169] as a binary splitter tree [151] with zero skew between the bits and stages, structured as an array of D flip flops. The clock signal from the last stage of the first register is bundled with a 64 bit data bus and connected to the second register. The clock signal, after some delay, is distributed within the second register by a similar H-tree network.

The circuit is schematically shown in Fig. 15.3, and WRspice [268] waveforms are depicted in Fig. 15.4. In this example, no additional area or energy savings from the proposed clocking scheme are achieved. The overall timing complexity is reduced since the clock signal of the receiving register is derived from the clock signal of the transmitting register. This dependent clock does not necessarily require zero skew relative to the primary clock source. By relaxing the requirements on the clocking system, the timing constraints on the EDA routing tools are also relaxed.

15.4.2 Behavior of H-Tree and Concurrent Networks

A 64 bit shift register with a depth of 8 bits and a concurrent clock distribution network are connected to one of the registers, as described in Sect. 15.4.1. The resulting topology is depicted in Fig. 15.5. The clock activation signal is generated by the AND-OR function between the clock signal of the transmitting block and the output data. This signal, connected through a delay line, is a concurrent clock signal within the receiving block.

Fig. 15.3 64 bit GALS clock
activation scheme for two
registers with an H-tree clock
distribution network

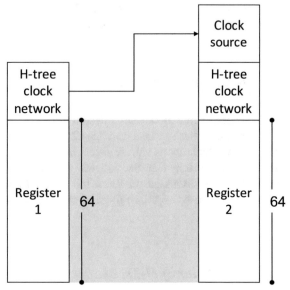

A simulation of this topology is depicted in Fig. 15.6. This circuit demonstrates the generation of a clock activation signal and a connection between two different clock distribution networks. In the case of a concurrent clock network within the receiving block, no additional clock source is needed as the clock signal travels together with the data packet.

15.4.3   Approaches for Clock Activation and Distribution

Two possible approaches exist for clock activation and distribution in the proposed scheme. The first approach is to distribute the activation signal through the clock distribution network. The resulting number of clock pulses precisely matches the number of incoming data pulses, and the clock is gated when data are not present. The benefit of this scheme is the absence of any activity when no data are being processed. The disadvantage, however, is careful control of the clock-data timing relationship is necessary [194]. The time to propagate the activation signal between blocks and through the clock distribution network needs to be less than the time for the data to propagate between the blocks plus one clock period. If this condition is not satisfied, fewer clock signals than data packets will be available. To satisfy this condition, additional delay may need to be added to the data path.

The second approach is to use the activation signal to turn on the source of the local clock. In this approach, the timing requirements are relaxed—the activation signal is only required to arrive before the arrival of the data packet. The disadvantage is that additional unnecessary clock pulses will be generated.

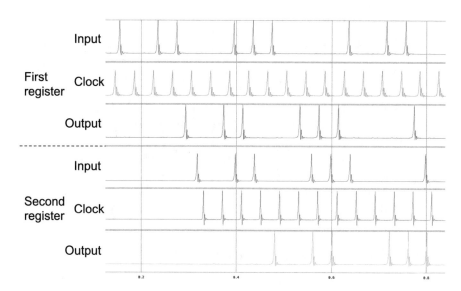

Fig. 15.4 Waveforms of a 1 bit slice of the circuit shown in Fig. 15.3

Fig. 15.5 64 bit GALS clock activation scheme for an H-tree clocked shift register connected to a shift register synchronized by a concurrent clock network

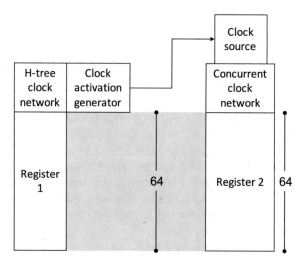

15.4.4 Multi-Chip Modules

Prospective SFQ systems will extensively utilize multi-chip modules (MCM) to improve integration and reduce the number of connections to a room temperature environment. Several functional blocks can be placed on different integrated circuits (ICs) within the same MCM. High speed communication between the ICs is possible through MCM bumps, where a data transmission rate of 93 GHz has

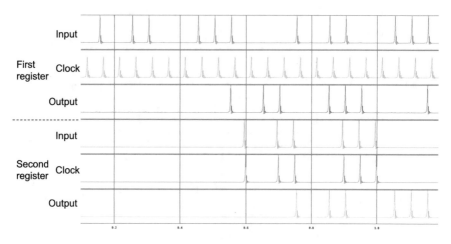

Fig. 15.6 Waveforms of a 1 bit slice of the circuit shown in Fig. 15.5

been demonstrated [391]. These connections are suitable for the proposed GALS technique, supporting prospective MCM systems.

15.4.5 Compatibility of Energy Efficient SFQ with Proposed Approaches

Energy efficient SFQ (ERSFQ) circuits utilize JTLs connected to the clock distribution network as a source of the highest average voltage within a circuit [50], as previously discussed in Sect. 4.3.2.4. As both the GALS clock activation scheme, described in Sect. 15.2, and the shared bus system, described in Sect. 15.3, gate the primary clock signal, current regulation within the energy efficient bias network is affected. To maintain compatibility with ERSFQ, the feeding JTLs of those circuits that utilize these proposed schemes are connected to an external average voltage reference [166].

Bias distribution for ERSFQ requires a power-up time—time required to distribute the bias currents. This time can be lowered by using smaller bias inductors or a larger feeding JTL [49]. Larger feeding JTLs with a lower bias inductance, which results in a faster power-up process, can be used in those circuits that can tolerate a power-up delay.

15.5 Conclusions

A globally asynchronous, locally synchronous clocking scheme based on clock activation signals and an intra-chip interconnect network for large scale SFQ systems are outlined in this chapter. The GALS clock activation scheme utilizes an enable signal generated by a functional block as a clock or a handshaking mechanism for communication with another block. The proposed GALS clock activation scheme reduces the complexity of the clock distribution network and provides timing flexibility and power savings with a routing area overhead of 3.1% for a 32 bit bus. These approaches combine existing CMOS asynchronous network-on-chip architectures, such as an intra-chip shared circular bus, with the unique features of SFQ circuits, thereby enabling high complexity SFQ VLSI and MCM-based systems.

Chapter 16
Design for Testability of SFQ Circuits

Test point insertion and set/scan techniques for enhanced testability in single flux quantum logic are discussed here. Test point insertion reduces the overhead of a set/scan chain while maintaining most of the functionality. Multiple ways of replacing costly (in terms of the number of Josephson junctions) SFQ multiplexers with mergers and blocking gates are presented. The multiplexer control signals are replaced with a gated clock signal or separate bias networks for both functional and test paths. Clocked blocking gates or current-controlled Josephson transmission line segments are used to disable undesirable data inputs. The clocked blocking gates for test point insertion in a 64 bit register require 35% fewer Josephson junctions as compared to multiplexers. This advantage further increases for current-controlled blocking gates. Set/scan chain and test point insertion techniques are applied to several SFQ circuits to evaluate error characteristics and to provide built-in self-test of SFQ-compatible memory systems [53, 54].

16.1 Introduction

Sub-terahertz clock frequencies achievable by SFQ circuits operating in a cryogenic environment make it difficult to generate and externally control test inputs via probing. Prototype testing of these circuits therefore requires advanced testing methodologies.

The testability of superconductive electronics, emphasizing defect-oriented and structural testing, has been previously considered [392]. Possible defects that can be introduced into a circuit during fabrication based on the physical layout and methods to detect these defects are reviewed [392]. An approach for built-in self-test (BIST) in RSFQ circuits was proposed [393]. In this chapter, a different approach is described to include a design for testability (DFT) capability for SFQ logic.

© Springer Nature Switzerland AG 2022
G. Krylov, E. G. Friedman, *Single Flux Quantum Integrated Circuit Design*,
https://doi.org/10.1007/978-3-030-76885-0_16

Testability features are introduced during the circuit design process, where these features are not dependent on the type of defect.

With the complexity and integration of conventional CMOS circuits, multiple automated testing techniques have been developed including BIST and automated test pattern generation (ATPG) [394]. A standard method to insert and control the test inputs and outputs produced by an arbitrarily complex circuit is the use of set/scan chain circuits [395]. Set/scan chains insert and observe information in serially connected flip flops [396]. The sequential circuits are disconnected from the combinatorial logic to form a long shift register through which test patterns are shifted by a clock signal (scan-in phase). After the state of a register is asserted, combinatorial logic is reconnected to produce an output state (capture phase). This output is read from the flip flops by connecting the flip flops into a shift register, and the data are shifted out of the register (scan-out phase). The data are compared with the expected output.

While the full scan approach provides observability and controllability to all of the flip flops in a chain, set/scan chains also introduce significant overhead. Each flip flop requires a multiplexer as well as additional area for routing. To reduce this overhead, a modified test methodology has been proposed for CMOS-based scan chains—test point insertion [397]. In this methodology, the combinatorial logic is included within the set/scan chain. The multiplexers are replaced with AND or OR gates and placed at the chain boundaries, greatly reducing the area overhead of the test structures.

Several notable differences between conventional CMOS logic and SFQ logic however exist, requiring standard CMOS-based DFT techniques to be modified to support SFQ logic. Traditional CMOS-based set/scan chains rely on a multiplexer to choose between normal operation and set/scan mode operation. In SFQ logic, however, a standard multiplexer requires 14 JJs for each bit [23] as compared to only 4 transistors in CMOS.

Unlike conventional CMOS logic, SFQ logic gates are inherently clocked and latched, as discussed in Chap. 5. Most SFQ logic gates consist of at least one storage loop. Moreover, each logic stage between sequentially adjacent registers [181] may require several clock cycles to produce an output. The test controller therefore needs to be aware of the number of cycles. Distinguishing between the combinatorial logic and the sequential blocks in SFQ logic is therefore more complex than in CMOS.

Another issue when applying set/scan chains to SFQ logic is the limited fanout of the gates and flip flops, as previously discussed in Chap. 4. The fanout of a standard SFQ logic gate and flip flop is one. Providing an additional output for a register requires a splitter cell (one splitter for each bit) [151]. The test outputs in a set/scan chain should therefore be placed sparingly to avoid this significant overhead.

While these issues are disadvantageous for the application of both full set/scan chains and test point insertion, SFQ logic also exhibits certain benefits. The power dissipated by SFQ circuits is extremely small. The additional DFT circuits therefore dissipate negligible power.

Test point insertion is therefore a preferable method when applying a black box approach to circuit testing, where only the input/output interfaces of a block are controllable. Set/scan chains are preferable when each logic stage is overly complex to ensure optimal coverage while only controlling the circuit interfaces.

In Sect. 16.2, several SFQ-specific modifications for DFT are presented. In Sects. 16.3 and 16.4, these improvements are applied, respectively, to the test point insertion technique and set/scan chains. In Sect. 16.5, the results are summarized.

16.2 Reducing DFT Overhead

The pulse-based nature of SFQ logic provides inherent isolation of the inputs from the different sources of data: if two outputs are connected to one via through a confluence buffer (CB), and no pulses arrive at one of the inputs (logic zero), the logic element functions correctly with only one input without requiring additional multiplexing circuitry. As a CB only requires 5 Josephson junctions as compared to 14 JJs for a typical multiplexer, a CB is preferable for test point operation. In the following subsections, several methods for replacing multiplexers with CBs are described.

16.2.1 Replacing Multiplexers

To replace the multiplexers with confluence buffers, it is necessary to ensure that no input pulses arrive at an inactive port of a CB. One method to ensure the absence of an input signal is to terminate any upstream signal generation by turning off the logic elements feeding data to the circuit under test (CUT). For the test path, this capability is achieved by a test controller, which supplies predefined test vectors during test mode and supplies logic zero during normal circuit operation. A regular logic path is, however, more difficult to turn off. When the controller initiates the test mode, the outputs of the upstream logic should be turned off to prevent these signals from affecting the test vectors. This capability requires test modes to disable the outputs from the upstream logic and introduces significant overhead. Furthermore, the test inputs cannot be inserted at arbitrary nodes within a logic path, and the test modes are not utilized during normal circuit operation.

Another method of disabling an undesired data path is to use special blocking gates that block an incoming signal or allow the signal to further propagate based on certain control signals. Multiple methods can be used to achieve this capability, where the choice depends upon the controller and circuit complexity. This approach, while also requiring considerable overhead, relaxes the requirements on the upstream logic, allowing a test input to be inserted at any point within a logic path.

16.2.2   Blocking Gates

In the following subsections, several possible forms of blocking gates are presented. These gates can be divided into two groups based on the control signal—clock-controlled or current-controlled.

16.2.2.1   Clock-Controlled Blocking Gate

A clocked blocking gate based on a D flip flop (DFF) is depicted in Fig. 16.1. This gate, essentially a part of a DFF, consists of a decision-making pair [398] combined with an auxiliary escape junction at the data input. All of the Josephson junctions are shunted, where the Stewart-McCumber parameter [399] β_c is set to 1. Depending upon the persistent current within the J3-L1-J1 loop, the arriving clock pulse either produces a 2π phase change in J2, leading to the absence of an SFQ pulse at the output, or triggers an output SFQ pulse by producing a 2π phase change in J1. Escape junction J3 prevents the blocking gate from being switched by two consecutive input pulses without a clock pulse. In this case, J3 initially switches, preventing a 2π phase change in J1. An output pulse is therefore not produced.

When the clock input is first enabled after the test mode is initiated, a single flux quantum may be present within the J3-L1-J1 loop. In this case, the flux quantum will escape and produce an output pulse which should be discarded. While this spurious output can be detrimental when the gate is used in logical operations, for DFT purposes, this pulse only extends the test mode by one clock cycle and therefore has a negligible effect on operation of the test mode.

The clocked blocking gate is combined with each bit of the normal data path and is enabled by a clock signal. This clock is gated during test mode. When the clock is enabled at this gate, the data path is transparent to any incoming SFQ pulses. When the clock is disabled, no output is produced. The gated clock train can be

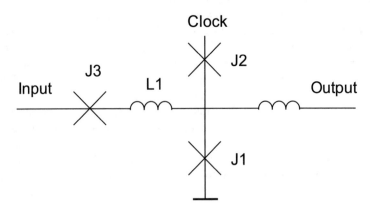

Fig. 16.1 DFF as a clocked blocking gate

produced by a nondestructive readout (NDRO) T flip flop attached to a clock output. Alternatively, an NDRO DFF can also be used. An NDRO T flip flop is shown in Fig. 16.2 and consists of a T flip flop and an NDRO D flip flop [23].

With a T flip flop toggle input, the circuit switches between the functional and test modes. A clock signal passes from a nearby register through an SFQ pulse splitter. The clock skew between the upstream logic and the blocking gate can be tuned to minimize the delay introduced by the blocking gate during normal operation [326, 400].

The temporal behavior of the combination of a clocked blocking gate with a merger is shown in Fig. 16.3. The circuit evaluated in this simulation is shown in Fig. 16.4. Correct multiplexing of data and test inputs is achieved, provided no signals originate from the test controller when not operating in the test mode. By attaching clocked blocking gates to both inputs of the merger, the full functionality of a multiplexer is duplicated independent of the data inputs while requiring only 11 JJs.

16.2.2.2   Current-Controlled Blocking Gates

Another possible method for a blocking gate is to reduce the bias current of the input JTL in front of the CB, thereby preventing propagation of an incoming SFQ pulse [47]. This method requires a controllable current, either provided by a source within the test controller or supplied externally. A blocking gate for this method is shown

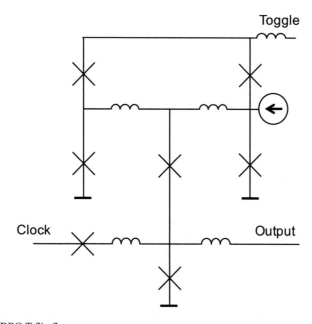

Fig. 16.2 NDRO T flip flop

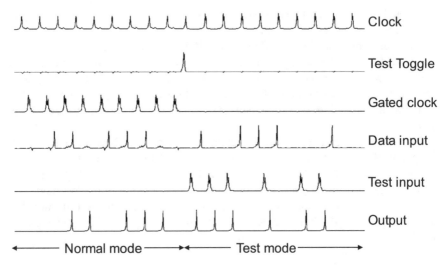

Fig. 16.3 Waveforms illustrating multiplexer operation performed by a blocking gate and a merger

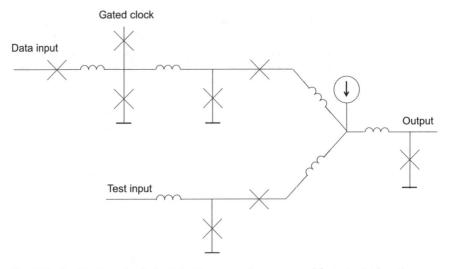

Fig. 16.4 Combination of a clocked blocking gate and a merger used for test point insertion

in Fig. 16.5. The current I_b controls the bias conditions of junction J2. When this current is low, J1 is closer to switching, allowing an incoming input pulse to switch J1, producing no output. When I_b is high, however, J2 is closer to switching, and the input pulse propagates along the path. The combination of two blocking gates and a CB, shown in Fig. 16.6, can multiplex two inputs depending upon the control current.

Waveforms describing the operation of this circuit are shown in Fig. 16.7. The primary benefit of this approach is greatly reduced area since no additional circuitry

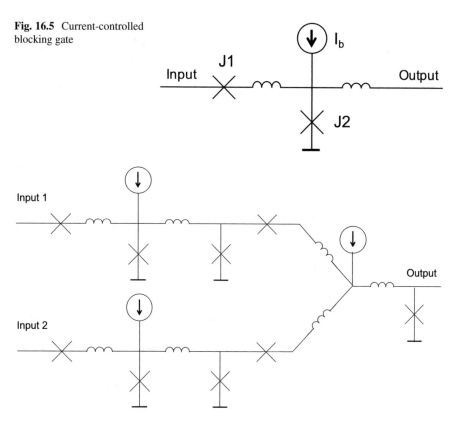

Fig. 16.5 Current-controlled blocking gate

Fig. 16.6 Multiplexer composed of two current-controlled blocking gates and a CB

is required within the chain other than the blocking gate shown in Fig. 16.5. Two separate bias distribution networks for both the functional and test paths with the ability to control the bias current are however required [49, 50]. The externally supplied current requires additional pins to control the different test points and set/scan chains [51]. Alternatively, the bias current can be generated and controlled internally by other SFQ circuits by supplying stored current from the NDRO flip flops to the blocking gates.

A similar method inductively couples the flux bias into the JTL input loop. An inductor inside a JTL segment is coupled to a controlled bias line [154]. When a current is supplied through this line, additional magnetic flux is introduced into the loop, which can, depending upon the direction, either prevent or assist the propagation of an incoming SFQ pulse by bringing either J1 or J2 closer to switching. The blocking gate for this method is depicted in Fig. 16.8.

In the following sections, both test point insertion and set/scan chain techniques are discussed. SFQ-specific methods for reducing the overhead are demonstrated on these two DFT techniques.

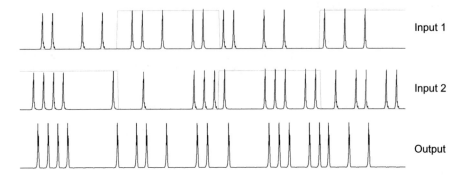

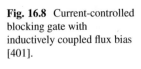

Fig. 16.7 Waveforms illustrating multiplexer operation performed by a current-controlled blocking gate and a CB. The bias current levels are shown as dotted lines overlapping the corresponding channel inputs

Fig. 16.8 Current-controlled blocking gate with inductively coupled flux bias [401].

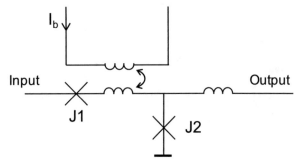

16.3   Test Point Insertion for SFQ Circuits

A modified test point insertion method targeted to SFQ circuits is described in this section. This method utilizes test inputs and outputs along the critical, most error-prone logic paths, or those logic paths where a black box testing approach is adequate. The test path contains both registers and logic gates, allowing correct operation of the combinatorial logic to be verified. By sparingly placing the test inputs and outputs, the overhead of the test circuits is reduced as compared to a set/scan chain. As these test points are inserted only at the input/output interfaces of the CUT, however, it is more difficult to modify the internal state of the CUT to achieve complete coverage.

The overall structure of the proposed method is schematically shown in Fig. 16.9. At the input, the CBs combine a normal data path with the test data path. Normal operation is disabled by a clocked blocking gate, as described in the following section. The test output drives SFQ pulse splitters at the output of the CUT.

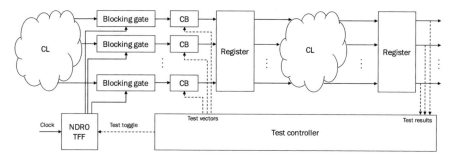

Fig. 16.9 Proposed test point insertion method. The dashed lines are active connections during test mode. A test point consists of blocking gates and a CB (similar to the circuits shown in Figs. 16.4 and 16.6) on every bit of a data path. In this figure, the blocking gates are supplied with a regular clock signal. This clock signal is gated by a NDRO T flip flop controlled by a test controller

16.3.1   Test Process

After blocking the normal data path, the CUT is supplied with test vectors. These vectors are chosen to provide sufficient coverage of the circuit functionality [396]. As each logic element requires multiple clock cycles to process the entire set/scan chain, the timing characteristics are included within the expected result by adding the correct number of clock cycles between output data.

The precise topology of the test controller largely depends upon the complexity of the CUT. The test controller can either include predetermined test vectors or the test vectors can be supplied by an external BIST/ATPG controller. Similarly, the correct output for the predetermined test vectors can either be stored on-chip in a ROM or stored externally. For current SFQ circuits, an on-chip controller is infeasible due to the large overhead. To reduce the number of output pins required by DFT in the case of an external controller, serial shift registers can load the test vectors and read out the results.

After the CUT produces a set of output responses for a set of input test vectors, the output response is passed to the test controller using splitter gates and transmission lines. These signals are compared with the expected results from the controller. Alternatively, the outputs can be compared off-chip.

16.3.2   Comparison of Blocking Gates with Multiplexers

Test point insertion in SFQ can be achieved with multiplexers. Blocking gates are compared here to a multiplexer. The number of JJs is used as a crude estimate of the total circuit area overhead, while clocking and routing overheads are assumed to be similar.

An SFQ multiplexer consists of 14 Josephson junctions switched by set/reset inputs [23]. The SFQ pulse mergers are combined with clocked blocking gates, requiring fewer JJs for the gates and only one control signal for the clock with log_2n splitters (as opposed to two networks, set and reset, required by a multiplexer).

The following expressions describe the number of JJs required for each test point as a function of the data bus width. The number of required junctions J_m if regular multiplexers are used is

$$J_m = (M + S) * n + 2S * log_2(n). \tag{16.1}$$

The number of required junctions J_c by the clocked blocking gates and confluence buffers is

$$J_c = (B_1 + C + S) * n + S * log_2(n) + T + S. \tag{16.2}$$

The number of required junctions J_b for the current-controlled blocking gates and confluence buffers is

$$J_b = (B_2 + C) * n, \tag{16.3}$$

where n is the width of a data path, M is the number of JJs in a multiplexer gate, and S and C are, respectively, the number of JJs in a splitter and confluence buffer. T is the number of JJs in a NDRO T flip flop. B_1 and B_2 are, respectively, the number of JJs in a clocked blocking gate segment and a current-controlled blocking gate. The number of JJs for each of these gates is listed in Table 16.1.

A comparison of the number of JJs required by the multiplexers and clocked blocking gates for increasing bus width is shown in Fig. 16.10. The advantages of the clocked blocking gates increase with wider data paths. For a 64 bit register, the use of multiplexers for test point insertion requires 1,124 JJs, while the use of clocked blocking gates requires only 732 JJs, 35% fewer JJs. With current-controlled blocking gates, this overhead is further reduced to only 448 JJs.

Table 16.1 Number of junctions per function [23]; see (16.1) to (16.6)

Parameter	Description	Number of JJs
M	Multiplexer	14
S	Splitter	3
B_1	Clocked blocking gate	3
B_2	Current-controlled blocking gate	2
C	Confluence buffer	5
T	NDRO T flip flop	7
I	Inverter	5

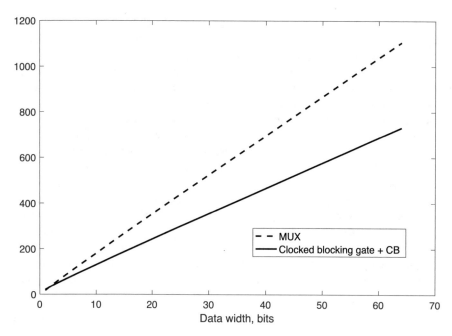

Fig. 16.10 Number of junctions in clocked blocking gates and confluence buffers as compared to multiplexers for different data widths

16.3.3   Advantages and Disadvantages of Test Point Insertion

As compared to the traditional set/scan chain approach widely used in CMOS, this technique does not place data in arbitrary registers, restricting the set of testable circuits. As there is no requirement for a circuit to be synchronous, however, this technique can also enhance testability in asynchronous and wave-pipelined systems [48, 52]. Multiple logic stages between registers are not bypassed; changes in the input data are therefore required to provide sufficient coverage. As this method requires sequential test pattern generation, similar to CMOS, considerable computational resources are used to choose an optimal set of test vectors and to determine the expected outputs, particularly with those logic paths that contain feedback [402].

16.4   Set/Scan Chains for SFQ Circuits

The test point insertion method can be logically extended to complete set/scan chains. As compared to test point insertion, a complete set/scan chain adds a separate path for the test data along the regular data path, requiring a multiplexer with each register. While this approach adds overhead and complexity, set/scan chains also provide numerous benefits: specifically, the ability to set and read each register in a chain and enhanced fault coverage of the logic gates, making set/scan chains effective in locating the precise location of an error.

Modifying the test point insertion process to achieve complete set/scan chains is shown in Fig. 16.11. For each logic block in a chain, an alternative path through a JTL is introduced. Regular SFQ multiplexers or a combination of blocking gates and CBs chooses between inputs. Blocking gates are placed on both the regular and test inputs of a CB, effectively forming a multiplexer. The blocking gates are switched by complementary control signals. Whenever one input is enabled, another input is disabled. In Fig. 16.11, clocked blocking gates are used as blocking gates to illustrate the additional circuitry required to produce the complementary clock signals. The clock signals are used as control signals to reduce the overhead as compared to separate control signals required by multiplexers. Clock distribution networks for clocked blocking gates can be reused by the combinatorial logic or flip flops, as shown in Fig. 16.11. If current-controlled blocking gates are used, the complementary clock distribution networks are replaced with two bias distribution networks—one network for the normal data path and one network for the test path.

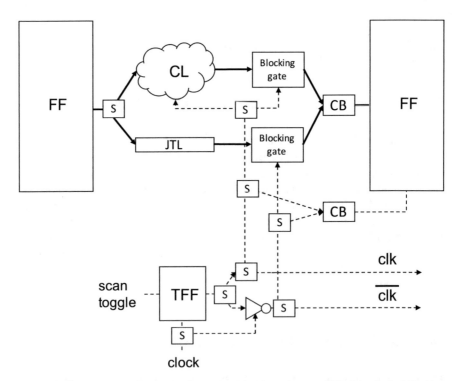

Fig. 16.11 Set/scan chain insertion method. The blocking gates are realized as clocked blocking gates. A T flip flop and inverter generate complementary clock signals, *clk* and $\overline{clk}$. Only one of the logic paths is active at any time—either regular mode (upper) or set/scan mode (lower). The data connections are shown as bold lines, and the control signals are shown as dashed lines

Similar to (16.1) to (16.3), each register in a set/scan chain using multiplexers requires J_m junctions, where J_m is a function of the data bus width,

$$J_m = (M + S) * n + 2S * log_2(n). \qquad (16.4)$$

For set/scan chains using blocking gates and confluence buffers, J_c junctions are required per register,

$$J_c = (2B_1 + C + S) * n + 2S * log_2(n) + 2S + T + I. \qquad (16.5)$$

And for scan chains using current controlled blocking gates and confluence buffers, J_b junctions per register are required,

$$J_b = (2B_2 + C) * n, \qquad (16.6)$$

where n is the width of the data path. The total number of junctions of a set/scan chain is scaled by the length of the chain. For this technique, current-controlled blocking gates further lower the overhead.

The primary issue with set/scan chains for SFQ is that each logic gate is typically clocked. Each logic block requires an arbitrary number of clock cycles to execute. While this behavior does not affect the scan-in and scan-out phases, the capture phase is further complicated. It is therefore desirable to insert set/scan chains into normal logic paths, where each stage requires the same number of clock cycles.

16.5 Conclusions

Test point insertion and set/scan chain techniques for SFQ circuits are presented in this chapter. These testability techniques are compatible with the RSFQ logic family and energy efficient modifications [33]. While both test point insertion and set/scan chains can use multiplexers, the number of required JJs can be greatly reduced by substituting multiplexers with confluence buffers and blocking gates.

Test point insertion can be used to evaluate critical and error-prone logic paths and is particularly useful for testing complex operations on wide data buses. In addition, this method can be used with a memory BIST controller with negligible overhead to provide access to the address, data, and control lines within a memory block. As the internal structure of the circuit is not modified, and synchronization is not required to operate correctly, test point insertion is also applicable to asynchronous and wave-pipelined circuits. Set/scan chains can be used for those circuits where fine control of the data within the registers is desired.

Chapter 17
Conclusions

Conventional semiconductor-based circuits have experienced major advances over the past few decades. Any further increase in the complexity of these deeply scaled circuits, however, faces many physical and engineering challenges, slowing the rate of increase in computational power. Once eclipsed by semiconductor circuits, superconductor-based circuits have recently emerged as a promising solution to these scaling issues. Superconductive circuits primarily target stationary applications in which the cost and inconvenience of cryogenic refrigeration can be offset by other benefits. The high computational efficiency of superconductive circuits, particularly, extremely low energy dissipation, is attractive for large scale data centers and cloud computing providers. The high speed and natural interface with superconductive qubits incentivize the use of these circuits in quantum computing systems. Other important applications of superconductive circuits include space-based systems where the ambient environment is naturally cryogenic.

The key to realizing the potential of superconductive electronics as a replacement for conventional circuits is scaling. Despite the many significant recent developments in SCE, superconductor-based circuits currently exhibit low complexity and are therefore limited to a few specific niche applications. In this book, the primary challenges of these circuits are introduced, among which are efficient memory, current biasing, clock distribution, and testability. Circuit and algorithmic approaches tackling these issues, while exploiting recent novel devices and advances in fabrication technology, are the focus of this book.

The lack of a fast and dense superconductive memory has long been a primary issue for superconductive systems. Emerging devices and fabrication techniques have the potential to mitigate this issue. Memory cells based on flux storage in SQUID loops can be scaled by using high inductance materials and JJs as inductive elements. Novel electrothermal, spintronic, and superconductor-ferromagnetic devices can function as storage and/or interface elements. Conventional and cryogenic CMOS memory, with drastically larger densities, can also be used within the memory hierarchy. The integration of disparate technologies within a single

© Springer Nature Switzerland AG 2022
G. Krylov, E. G. Friedman, *Single Flux Quantum Integrated Circuit Design*,
https://doi.org/10.1007/978-3-030-76885-0_17

IC, multi-chip module, or overall memory stack is challenging. Tradeoffs among performance, area, power, thermal characteristics, and integration capability require careful consideration.

The clock distribution network is a primary component of any digital IC, providing a temporal reference for synchronous communication. Pulse-based logic, such as RSFQ, is particularly challenging for the clock tree synthesis process. The primary distinctive features of these circuits—clocked logic gates and multi-gigahertz clock frequencies—drastically increase the number of clock sinks while narrowing timing tolerances. Automated clock tree synthesis algorithms have been developed to construct a clock tree suitable for complex superconductive systems. AC clocked logic, such as RQL or AQFP, utilizes a multiphase AC clock, where the phase differences among the gates are carefully tuned. Modern self-timed and dynamic SFQ circuits significantly reduce the size and timing requirements of the clock network. Globally asynchronous, locally synchronous clocking schemes help to manage and constrain overall complexity. These novel synchronization techniques and algorithms are used to further exploit the benefits of superconductive circuits.

Power generation and distribution is another challenge inherent to large scale ICs. DC-biased superconductive circuits require large bias currents, which are difficult to supply and distribute. Current recycling techniques can be used to reduce the total bias current in DC-biased logic families. Near zero static leakage current is an important advantage contributing to the power efficiency of superconductive circuits. To realize this potential, modern bias current delivery systems exploit inductive bias distribution networks. Design guidelines and algorithms for synthesizing bias networks utilize these novel features and provide effective ways for placing the necessary circuit components.

On-chip testability is necessary to support the automated evaluation of complex ICs. In superconductive circuits, the need for integrated design for testability features is exacerbated by the high clock frequencies and cryogenic operation, preventing or complicating many forms of manual testing. Many testability techniques commonly used in conventional semiconductor circuits, such as set/scan chains, have been adapted for SCE. Novel circuit features are required to enhance the observability and controllability of individual blocks or to provide a means to locally initialize or reset a circuit.

The topics presented in this book are intended to review the current state of development of superconductive digital logic while providing a physical perspective and engineering insight into the many challenges faced by this technology. Methodologies and techniques targeting large scale integration of superconductive systems are crucial to continue the development and scaling of these circuits. The performance benefits and energy efficiency of superconductive circuits, applied to important mainstream as well as evolving applications, can transform superconductive electronics into an essential technology for large scale, stationary integrated computing systems.

Bibliography

1. H. K. Onnes, "Investigations into the Properties of Substances at Low Temperatures, Which Have Led, Amongst Other Things, to the Preparation of Liquid Helium," *Nobel Lecture*, vol. 4, December 1913.
2. T. Jenkins, "A Brief History of... Semiconductors," *Physics Education*, vol. 40, no. 5, p. 430, February 2005.
3. J. Bardeen and W. H. Brattain, "The Transistor, a Semi-Conductor Triode," *Physical Review*, vol. 74, no. 2, p. 230, July 1948.
4. L. J. Lilienfeld, "Method and Apparatus for Controlling Electric Currents," U.S. Patent No. 1,745,175A, October 22, 1925.
5. D. A. Buck, "The Cryotron – a Superconductive Computer Component," *Proceedings of the IRE*, vol. 44, no. 4, pp. 482–493, April 1956.
6. K. K. Likharev, "Superconductor Digital Electronics," *Physica C: Superconductivity and its Applications*, vol. 482, pp. 6–18, November 2012.
7. "Cryotrons May Lead to Computers Cubic Foot in Size," *The New York Times*, p. 1, February 6, 1957.
8. J. S. Kilby, "Miniaturized Electronic Circuits," U.S. Patent No. 3,138,743, June 23, 1964.
9. J. Bremer, "The Invention of Superconducting Integrated Circuit," *IEEE History Center Newsletter*, vol. 75, pp. 6–7, November 2007.
10. D. L. Shell, "The Share 709 System: A Cooperative Effort," *Journal of the ACM*, vol. 6, no. 2, p. 123–127, April 1959.
11. J. Raymond and D. K. Banerji, "Using a Microprocessor in an Intelligent Graphics Terminal," *Computer*, vol. 9, no. 4, pp. 18–25, April 1976.
12. G. E. Moore, "Cramming More Components onto Integrated Circuits," *Electronics*, vol. 38, no. 8, p. 114–117, April 1965.
13. B. D. Josephson, "Possible New Effects in Superconductive Tunnelling," *Physics Letters*, vol. 1, no. 7, pp. 251–253, July 1962.
14. H. H. Zappe and K. R. Grebe, "Dynamic Behavior of Josephson Tunnel Junctions in the Subnanosecond Range," *Journal of Applied Physics*, vol. 44, no. 2, pp. 865–874, February 1973.
15. R. L. Van Tuyl, C. A. Liechti, R. E. Lee, and E. Gowen, "GaAs MESFET logic with 4-GHz clock rate," *IEEE Journal of Solid-State Circuits*, vol. 12, no. 5, pp. 485–496, October 1977.
16. L. Esaki, "New Phenomenon in Narrow Germanium $p - n$ Junctions," *Physical Review*, vol. 109, pp. 603–604, January 1958.
17. Y. Hazoni, "A Fast Flip-Flop Circuit Utilizing Tunnel-Diodes," *Nuclear Instruments & Methods*, vol. 13, pp. 95–96, August-October 1961.

© Springer Nature Switzerland AG 2022
G. Krylov, E. G. Friedman, *Single Flux Quantum Integrated Circuit Design*,
https://doi.org/10.1007/978-3-030-76885-0

18. W. J. Gallagher, E. P. Harris, and M. B. Ketchen, "Superconductivity at IBM – A Centennial Review: Part I – Superconducting Computer and Device Applications," *Proceedings of the IEEE/CSC ESAS European Superconductivity News Forum*, no. 21, pp. 1–34, July 2012.

19. H. Nakagawa, I. Kurosawa, M. Aoyagi, S. Kosaka, Y. Hamazaki, Y. Okada, and S. Takada, "A 4-bit Josephson Computer ETL-JC1," *IEEE Transactions on Applied Superconductivity*, vol. 1, no. 1, pp. 37–47, March 1991.

20. D. C. Brock, "Will the NSA Finally Build Its Superconducting Spy Computer?" *IEEE Spectrum*, February 2016.

21. J. G. Bednorz and K. A. Müller, "Possible High T_c Superconductivity in the Ba–La–Cu–O System," *Zeitschrift für Physik B Condensed Matter*, vol. 64, no. 2, pp. 189–193, June 1986.

22. H. Toepfer, T. Ortlepp, H. F. Uhlmann, D. Cassel, and M. Siegel, "Design of HTS RSFQ circuits," *Physica C: Superconductivity*, vol. 392-396, pp. 1420–1425, October 2003.

23. K. K. Likharev and V. K. Semenov, "RSFQ Logic/Memory Family: a New Josephson-Junction Technology for Sub-Terahertz-Clock-Frequency Digital Systems," *IEEE Transactions on Applied Superconductivity*, vol. 1, no. 1, pp. 3–28, March 1991.

24. M. Hosoya, W. Hioe, J. Casas, R. Kamikawai, Y. Harada, Y. Wada, H. Nakane, R. Suda, and E. Goto, "Quantum Flux Parametron: a Single Quantum Flux Device for Josephson Supercomputer," *IEEE Transactions on Applied Superconductivity*, vol. 1, no. 2, pp. 77–89, June 1991.

25. D. S. Holmes, A. L. Ripple, and M. A. Manheimer, "Energy-Efficient Superconducting Computing – Power Budgets and Requirements," *IEEE Transactions on Applied Superconductivity*, vol. 23, no. 3, p. 1701610, February 2013.

26. M. A. Manheimer, "Cryogenic Computing Complexity Program: Phase 1 Introduction," *IEEE Transactions on Applied Superconductivity*, vol. 25, no. 3, pp. 1–4, June 2015.

27. R. K. Cavin, P. Lugli, and V. V. Zhirnov, "Science and Engineering Beyond Moore's Law," *Proceedings of the IEEE*, vol. 100, no. Special Centennial Issue, pp. 1720–1749, May 2012.

28. H. Esmaeilzadeh, E. Blem, R. S. Amant, K. Sankaralingam, and D. Burger, "Dark Silicon and the End of Multicore Scaling," *Proceedings of the ACM/IEEE Annual International Symposium on Computer Architecture*, pp. 365–376, June 2011.

29. J. A. Hutchby, G. I. Bourianoff, V. V. Zhirnov, and J. E. Brewer, "Extending the Road Beyond CMOS," *IEEE Circuits and Devices Magazine*, vol. 18, no. 2, pp. 28–41, August 2002.

30. V. V. Dotsenko *et al.*, "Integrated Cryogenic Electronics Testbed (ICE-T) for Evaluation of Superconductor and Cryo-Semiconductor Integrated Circuits," *IOP Conference Series: Materials Science and Engineering*, vol. 171, p. 012145, February 2017.

31. J. Ekin, *Experimental Techniques for Low-Temperature Measurements: Cryostat Design, Material Properties and Superconductor Critical-Current Testing*, Oxford University Press, 2006.

32. R. Patterson, A. Hammoud, and M. Elbuluk, "Assessment of Electronics for Cryogenic Space Exploration Missions," *Cryogenics*, vol. 46, no. 2-3, pp. 231–236, February/March 2006.

33. O. A. Mukhanov, "Energy-Efficient Single Flux Quantum Technology," *IEEE Transactions on Applied Superconductivity*, vol. 21, no. 3, pp. 760–769, June 2011.

34. W. Chen, A. V. Rylyakov, V. Patel, J. E. Lukens, and K. K. Likharev, "Rapid Single Flux Quantum T-Flip Flop Operating up to 770 GHz," *IEEE Transactions on Applied Superconductivity*, vol. 9, no. 2, pp. 3212–3215, June 1999.

35. O. A. Mukhanov, D. Gupta, A. M. Kadin, and V. K. Semenov, "Superconductor Analog-to-Digital Converters," *Proceedings of the IEEE*, vol. 92, no. 10, pp. 1564–1584, October 2004.

36. N. Takeuchi, Y. Yamanashi, and N. Yoshikawa, "Reversible Logic Gate using Adiabatic Superconducting Devices," *Scientific Reports*, vol. 4, p. 6354, September 2014.

37. J. M. Lockhart, "SQUID Readout and Ultra-Low Magnetic Fields for Gravity Probe-B (GP-B)," *Proceedings of the SPIE, Cryogenic Optical Systems and Instruments II*, vol. 619, pp. 148–156, July 1986.

38. G. N. Gol'tsman, O. Okunev, G. Chulkova, A. Lipatov, A. Semenov, K. Smirnov, B. Voronov, A. Dzardanov, C. Williams, and R. Sobolewski, "Picosecond Superconducting Single-Photon Optical Detector," *Applied Physics Letters*, vol. 79, no. 6, pp. 705–707, August 2001.

39. R. McDermott, M. G. Vavilov, B. L. T. Plourde, F. K. Wilhelm, P. J. Liebermann, O. A. Mukhanov, and T. A. Ohki, "Quantum–Classical Interface based on Single Flux Quantum Digital Logic," *Quantum Science and Technology*, vol. 3, no. 2, p. 024004, January 2018.

40. U. Ghoshal and T. Van Duzer, "Superconductivity Researchers Seek to Remove Computational Bottlenecks: Wide Communication Bandwidths and Fast Switching Make Superconductive Technology Look Attractive in Computer Applications," *Computers in Physics*, vol. 6, no. 6, pp. 585–593, November/December 1992.

41. D. E. Nikonov and I. A. Young, "Benchmarking of Beyond-CMOS Exploratory Devices for Logic Integrated Circuits," *IEEE Journal on Exploratory Solid-State Computational Devices and Circuits*, vol. 1, pp. 3–11, April 2015.

42. A. Gara *et al.*, "Overview of the Blue Gene/L System Architecture," *IBM Journal of Research and Development*, vol. 49, no. 2/3, pp. 195–212, March/May 2005.

43. S. Krinner, S. Storz, P. Kurpiers, P. Magnard, J. Heinsoo, R. Keller, J. Lütolf, C. Eichler, and A. Wallraff, "Engineering Cryogenic Setups for 100-Qubit Scale Superconducting Circuit Systems," *EPJ Quantum Technology*, vol. 6, no. 1, May 2019.

44. G. Krylov, J. Kawa, and E. G. Friedman, "Design Automation of Superconductive Digital Circuits: A Review," *IEEE Nanotechnology Magazine*, December 2021 (in press).

45. G. Krylov and E. G. Friedman, "Behavioral Verilog-A Model of Superconductor-Ferromagnetic Transistor," *Proceedings of the IEEE International Symposium on Circuits and Systems*, May 2018.

46. G. Krylov and E. G. Friedman, "Inductive Noise Coupling in Superconductive Passive Transmission Lines," *Proceedings of the IEEE International Midwest Symposium on Circuits and Systems*, pp. 727–731, August 2021.

47. G. Krylov and E. G. Friedman, "Sense Amplifier for Spin-Based Cryogenic Memory Cells," *IEEE Transactions on Applied Superconductivity*, vol. 29, no. 5, pp. 1–4, Art no. 1 501 804, August 2019.

48. G. Krylov and E. G. Friedman, "Asynchronous Dynamic Single Flux Quantum Majority Gates," *IEEE Transactions on Applied Superconductivity*, vol. 30, no. 5, pp. 1–7, Art no. 1 300 907, August 2020.

49. G. Krylov and E. G. Friedman, "Bias Distribution in ERSFQ VLSI Circuits," *Proceedings of the IEEE International Symposium on Circuits and Systems*, pp. 1–5, October 2020.

50. G. Krylov and E. G. Friedman, "Design Methodology for Distributed Large Scale ERSFQ Bias Networks," *IEEE Transactions on Very Large Scale Integration (VLSI) Systems*, vol. 28, no. 11, pp. 2438–2447, November 2020.

51. G. Krylov and E. G. Friedman, "Partitioning RSFQ Circuits for Current Recycling," *IEEE Transactions on Applied Superconductivity*, vol. 31, no. 5, pp. 1–6, Art no. 1 301 706, August 2021.

52. G. Krylov and E. G. Friedman, "Globally Asynchronous, Locally Synchronous Clocking and Shared Interconnect for Large-Scale SFQ Systems," *IEEE Transactions on Applied Superconductivity*, vol. 29, no. 5, pp. 1–5, Art no. 3 603 205, August 2019.

53. G. Krylov and E. G. Friedman, "Test Point Insertion for RSFQ Circuits," *Proceedings of the IEEE International Symposium on Circuits and Systems*, pp. 2022–2025, May 2017.

54. G. Krylov and E. G. Friedman, "Design for Testability of SFQ Circuits," *IEEE Transactions on Applied Superconductivity*, vol. 27, no. 8, pp. 1–7, Art no. 1 302 307, December 2017.

55. F. London and H. London, "The Electromagnetic Equations of the Supraconductor," *Proceedings of the Royal Society of London. Series A – Mathematical and Physical Sciences*, vol. 149, no. 866, pp. 71–88, March 1935.

56. V. L. Ginzburg, "On Superconductivity and Superfluidity," *Nobel Lecture*, December 2003.

57. P. C. Hohenberg and A. P. Krekhov, "An Introduction to the Ginzburg–Landau Theory of Phase Transitions and Nonequilibrium Patterns," *Physics Reports*, vol. 572, pp. 1–42, April 2015.

58. V. L. Ginzburg, "Superfluidity and Superconductivity in Astrophysics," *Comments on Astrophysics and Space Physics*, vol. 1, pp. 81–86, May 1969.

59. J. Bardeen, L. N. Cooper, and J. R. Schrieffer, "Theory of Superconductivity," *Physical Review*, vol. 108, no. 5, pp. 1175–1204, December 1957.

60. L. N. Cooper, "Bound Electron Pairs in a Degenerate Fermi Gas," *Physical Review*, vol. 104, pp. 1189–1190, November 1956.

61. V. F. Weisskopf, *The Formation of Cooper Pairs and the Nature of Superconducting Currents*, CERN, Switzerland, 1979.

62. B. V. Svistunov, E. S. Babaev, and N. V. Prokof'ev, *Superfluid States of Matter*, CRC Press, 2015.

63. W. Meissner and R. Ochsenfeld, "Ein Neuer Effekt bei Eintritt der Supraleitfähigkeit," *Naturwissenschaften*, vol. 21, no. 44, pp. 787–788, November 1933.

64. L. H. Greene, "High-Temperature Superconductors: Playgrounds for Broken Symmetries," *AIP Conference Proceedings*, vol. 795, no. 1, pp. 70–82, October 2005.

65. S. A. Kivelson and D. S. Rokhsar, "Bogoliubov Quasiparticles, Spinons, and Spin-Charge Decoupling in Superconductors," *Physical Review B*, vol. 41, pp. 11 693–11 696, June 1990.

66. A. I. Golovashkin and N. P. Shabanova, "Temperature Dependence of Critical Magnetic Fields and Electronic Characteristics of Nb3Ge Films," *Soviet Physics, JETP*, vol. 55, no. 3, pp. 503–508, March 1982.

67. A. B. Pippard, "Field Variation of the Superconducting Penetration Depth," *Proceedings of the Royal Society of London. Series A. Mathematical and Physical Sciences*, vol. 203, no. 1073, pp. 210–223, May 1950.

68. J. E. Sonier, *The Magnetic Penetration Depth and the Vortex Core Radius in Type-II Superconductors*, Ph.D. Dissertation, University of British Columbia, Vancouver, Canada, 1998.

69. A. A. Abrikosov, "Nobel Lecture: Type-II Superconductors and the Vortex Lattice," *Reviews of Modern Physics*, vol. 76, pp. 975–979, December 2004.

70. R. A. French, "Intrinsic Type-2 Superconductivity in Pure Niobium," *Cryogenics*, vol. 8, no. 5, pp. 301–308, October 1968.

71. H. A. Boorse, D. B. Cook, and M. W. Zemansky, "Superconductivity of Lead," *Physical Review*, vol. 78, pp. 635–636, June 1950.

72. I. S. Khukhareva, "The Superconducting Properties of Thin Aluminum Films," *Soviet Physics, JETP*, vol. 16, pp. 828–832, April 1963.

73. G. Behrens, W. Campbell, D. Williams, and S. White, "Guidelines for the Design of Cryogenic Systems," *NRAO Electronic Division Internal Report*, no. 306, 1997.

74. A. Schilling, M. Cantoni, J. Guo, and H. Ott, "Superconductivity above 130 K in the Hg–Ba–Ca–Cu–O System," *Nature*, vol. 363, no. 6424, pp. 56–58, May 1993.

75. K. K. Likharev, *Dynamics of Josephson Junctions and Circuits*, Gordon and Breach Science Publishers, 1986.

76. J. Clarke, "Supercurrents in Lead-Copper-Lead Sandwiches," *Proceedings of the Royal Society of London. Series A. Mathematical and Physical Sciences*, vol. 308, no. 1495, pp. 447–471, January 1969.

77. K. K. Likharev, "Superconducting Weak Links," *Reviews of Modern Physics*, vol. 51, pp. 101–159, January 1979.

78. S. K. Tolpygo, "Superconductor Digital Electronics: Scalability and Energy Efficiency Issues," *Low Temperature Physics*, vol. 42, no. 5, pp. 361–379, June 2016.

79. R. Gross, A. Marx, and F. Deppe, *Applied Superconductivity: Josephson Effect and Superconducting Electronics*, Walter De Gruyter Incorporated, 2016.

80. W. C. Stewart, "Current-Voltage Characteristics of Josephson Junctions," *Applied Physics Letters*, vol. 12, no. 8, pp. 277–280, April 1968.

81. D. E. McCumber, "Effect of AC Impedance on DC Voltage-Current Characteristics of Superconductor Weak-Link Junctions," *Journal of Applied Physics*, vol. 39, no. 7, pp. 3113–3118, June 1968.

82. E. E. Wollman, V. B. Verma, A. E. Lita, W. H. Farr, M. D. Shaw, R. P. Mirin, and S. W. Nam, "Kilopixel Array of Superconducting Nanowire Single-Photon Detectors," *Optics Express*, vol. 27, no. 24, pp. 35 279–35 289, November 2019.

83. C. M. Natarajan, M. G. Tanner, and R. H. Hadfield, "Superconducting Nanowire Single-Photon Detectors: Physics and Applications," *Superconductor Science and Technology*, vol. 25, no. 6, p. 063001, April 2012.

84. A. N. McCaughan and K. K. Berggren, "A Superconducting-Nanowire Three-Terminal Electrothermal Device," *Nano Letters*, vol. 14, no. 10, pp. 5748–5753, September 2014.

85. I. P. Nevirkovets, O. Chernyashevskyy, G. V. Prokopenko, O. A. Mukhanov, and J. B. Ketterson, "Control of Supercurrent in Hybrid Superconducting–Ferromagnetic Transistors," *IEEE Transactions on Applied Superconductivity*, vol. 25, no. 3, pp. 1–5, June 2015.

86. I. P. Nevirkovets, O. Chernyashevskyy, G. V. Prokopenko, O. A. Mukhanov, and J. B. Ketterson, "Superconducting-Ferromagnetic Transistor," *IEEE Transactions on Applied Superconductivity*, vol. 24, no. 4, pp. 1–6, August 2014.

87. S. Faris, S. Raider, W. Gallagher, and R. Drake, "Quiteron," *IEEE Transactions on Magnetics*, vol. 19, no. 3, pp. 1293–1295, May 1983.

88. S. Shafranjuk, I. Nevirkovets, O. Mukhanov, and J. Ketterson, "Control of Superconductivity in a Hybrid Superconducting/Ferromagnetic Multilayer using Nonequilibrium Tunneling Injection," *Physical Review Applied*, vol. 6, no. 2, p. 024018, August 2016.

89. A. I. Buzdin, "Proximity Effects in Superconductor-Ferromagnet Heterostructures," *Reviews of Modern Physics*, vol. 77, no. 3, pp. 935–976, September 2005.

90. J. Matisoo, "The Tunneling Cryotron – A Superconductive Logic Element Based on Electron Tunneling," *Proceedings of the IEEE*, vol. 55, no. 2, pp. 172–180, February 1967.

91. K. Likharev, O. Mukhanov, and V. Semenov, "Resistive Single Flux Quantum Logic for the Josephson-Junction Digital Technology," *Proceedings of the Third International Conference on Superconducting Quantum Devices*, pp. 1103–1108, June 1985.

92. Q. P. Herr, A. Y. Herr, O. T. Oberg, and A. G. Ioannidis, "Ultra-Low-Power Superconductor Logic," *Journal of Applied Physics*, vol. 109, no. 10, p. 103903, May 2011.

93. S. Kumar, W. F. Avrin, and B. R. Whitecotton, "NMR of Room Temperature Samples with a Flux-Locked DC SQUID," *IEEE Transactions on Magnetics*, vol. 32, no. 6, pp. 5261–5264, November 1996.

94. A. H. Silver and J. E. Zimmerman, "Quantum Transitions and Loss in Multiply Connected Superconductors," *Physical Review Letters*, vol. 15, pp. 888–891, December 1965.

95. T. Van Duzer and C. W. Turner, *Principles of Superconductive Devices and Circuits*, Prentice Hall, 1999.

96. K. Likharev, "Dynamics of Some Single Flux Quantum Devices: I. Parametric Quantron," *IEEE Transactions on Magnetics*, vol. 13, no. 1, pp. 242–244, January 1977.

97. R. L. Fagaly, "Superconducting Quantum Interference Device Instruments and Applications," *Review of Scientific Instruments*, vol. 77, no. 10, p. 101101, October 2006.

98. R. Jaklevic, J. Lambe, A. Silver, and J. Mercereau, "Quantum Interference Effects in Josephson Tunneling," *Physical Review Letters*, vol. 12, no. 7, p. 159, February 1964.

99. T. R. Gheewala, "A 30-ps Josephson Current Injection Logic (CIL)," *IEEE Journal of Solid-State Circuits*, vol. 14, no. 5, pp. 787–793, October 1979.

100. H. H. Zappe, "A Single Flux Quantum Josephson Junction Memory Cell," *Applied Physics Letters*, vol. 25, no. 7, pp. 424–426, June 1974.

101. E. Salman and Eby G. Friedman, *High Performance Integrated Circuit Design*, McGraw-Hill Publishers, 2012.

102. T. R. Gheewala, "Josephson Logic Circuits based on Nonlinear Current Injection in Interferometer Devices," *Applied Physics Letters*, vol. 33, no. 8, pp. 781–783, October 1978.

103. M. Klein and D. J. Herrell, "Sub-100 ps Experimental Josephson Interferometer Logic Gates," *IEEE Journal of Solid-State Circuits*, vol. 13, no. 5, pp. 577–583, October 1978.

104. T. Fulton and R. Dynes, "Switching to Zero Voltage in Josephson Tunnel Junctions," *Solid State Communications*, vol. 9, no. 13, pp. 1069–1073, July 1971.

105. R. Jewett and T. Van Duzer, "Low-Probability Punchthrough in Josephson Junctions," *IEEE Transactions on Magnetics*, vol. 17, no. 1, pp. 599–602, January 1981.

106. R. L. Van Tuyl and C. A. Liechti, "High-Speed Integrated Logic with GaAs MESFET's," *IEEE Journal of Solid-State Circuits*, vol. 9, no. 5, pp. 269–276, October 1974.

107. T. Jabbari, G. Krylov, S. Whiteley, E. Mlinar, J. Kawa, and E. G. Friedman, "Interconnect Routing for Large-Scale RSFQ Circuits," *IEEE Transactions on Applied Superconductivity*, vol. 29, no. 5, pp. 1–5, Art no. 1 102 805, August 2019.

108. O. T. Oberg, *Superconducting Logic Circuits Operating with Reciprocal Magnetic Flux Quanta*, Ph.D. Dissertation, University of Maryland, College Park, Maryland, 2011.

109. I. I. Soloviev, N. V. Klenov, S. V. Bakurskiy, M. Y. Kupriyanov, A. L. Gudkov, and A. S. Sidorenko, "Beyond Moore's Technologies: Operation Principles of a Superconductor Alternative," *Beilstein Journal of Nanotechnology*, vol. 8, pp. 2689–2710, November 2017.

110. A. M. Kadin, R. J. Webber, and S. Sarwana, "Effects of Superconducting Return Currents on RSFQ Circuit Performance," *IEEE Transactions on Applied Superconductivity*, vol. 15, no. 2, pp. 280–283, June 2005.

111. A. Y. Herr, Q. P. Herr, O. T. Oberg, O. Naaman, J. X. Przybysz, P. Borodulin, and S. B. Shauck, "An 8-bit Carry Look-Ahead Adder with 150 ps Latency and Sub-Microwatt Power Dissipation at 10 GHz," *Journal of Applied Physics*, vol. 113, no. 3, p. 033911, January 2013.

112. V. K. Semenov, Y. A. Polyakov, and S. K. Tolpygo, "Very Large Scale Integration of Josephson-Junction-Based Superconductor Random Access Memories," *IEEE Transactions on Applied Superconductivity*, vol. 29, no. 5, pp. 1–9, August 2019.

113. R. Landauer, "Irreversibility and Heat Generation in the Computing Process," *IBM Journal of Research and Development*, vol. 5, no. 3, pp. 183–191, July 1961.

114. J. P. Peterson, R. S. Sarthour, A. M. Souza, I. S. Oliveira, J. Goold, K. Modi, D. O. Soares-Pinto, and L. C. Céleri, "Experimental Demonstration of Information to Energy Conversion in a Quantum System at the Landauer Limit," *Proceedings of the Royal Society A: Mathematical, Physical and Engineering Sciences*, vol. 472, no. 2188, p. 20150813, April 2016.

115. C. H. Bennett, "Logical Reversibility of Computation," *IBM Journal of Research and Development*, vol. 17, no. 6, pp. 525–532, November 1973.

116. Y. Harada, H. Nakane, N. Miyamoto, U. Kawabe, E. Goto, and T. Soma, "Basic Operations of the Quantum Flux Parametron," *IEEE Transactions on Magnetics*, vol. 23, no. 5, pp. 3801–3807, September 1987.

117. N. Takeuchi, D. Ozawa, Y. Yamanashi, and N. Yoshikawa, "An Adiabatic Quantum Flux Parametron as an Ultra-Low-Power Logic Device," *Superconductor Science and Technology*, vol. 26, no. 3, p. 035010, January 2013.

118. O. Chen, R. Cai, Y. Wang, F. Ke, T. Yamae, R. Saito, N. Takeuchi, and N. Yoshikawa, "Adiabatic Quantum-Flux-Parametron: Towards Building Extremely Energy-Efficient Circuits and Systems," *Scientific Reports*, vol. 9, no. 1, pp. 1–10, July 2019.

119. T. Clark and J. Baldwin, "Superconducting Memory Device Using Josephson Junctions," *Electronics Letters*, vol. 3, no. 5, pp. 178–179, May 1967.

120. S. Tahara, I. Ishida, Y. Ajisawa, and Y. Wada, "Experimental Vortex Transitional Nondestructive Read-Out Josephson Memory Cell," *Journal of Applied Physics*, vol. 65, no. 2, pp. 851–856, January 1989.

121. S. K. Tolpygo, V. Bolkhovsky, D. E. Oates, R. Rastogi, S. Zarr, A. L. Day, T. J. Weir, A. Wynn, and L. M. Johnson, "Superconductor Electronics Fabrication Process with MoNx Kinetic Inductors and Self-Shunted Josephson Junctions," *IEEE Transactions on Applied Superconductivity*, vol. 28, no. 4, pp. 1–12, June 2018.

122. Y. Kim *et al.*, "A 16-Gb, 18-Gb/s/pin GDDR6 DRAM With Per-Bit Trainable Single-Ended DFE and PLL-Less Clocking," *IEEE Journal of Solid-State Circuits*, vol. 54, no. 1, pp. 197–209, January 2019.

123. N. Yoshikawa, T. Tomida, M. Tokuda, Q. Liu, X. Meng, S. R. Whiteley, and T. Van Duzer, "Characterization of 4 K CMOS Devices and Circuits for Hybrid Josephson-CMOS Systems," *IEEE Transactions on Applied Superconductivity*, vol. 15, no. 2, pp. 267–271, June 2005.

124. W. F. Clark, B. El-Kareh, R. G. Pires, S. L. Titcomb, and R. L. Anderson, "Low Temperature CMOS - a Brief Review," *IEEE Transactions on Components, Hybrids, and Manufacturing Technology*, vol. 15, no. 3, pp. 397–404, June 1992.

125. H. Suzuki, A. Inoue, T. Imamura, and S. Hasuo, "A Josephson Driver to Interface Josephson Junctions to Semiconductor Transistors," *Proceedings of the IEEE International Electron Devices Meeting*, pp. 290–293, December 1988.

126. T. Ortlepp, S. R. Whiteley, L. Zheng, X. Meng, and T. Van Duzer, "High-Speed Hybrid Superconductor-to-Semiconductor Interface Circuit With Ultra-Low Power Consumption," *IEEE Transactions on Applied Superconductivity*, vol. 23, no. 3, p. 1400104, June 2013.

127. T. Van Duzer, L. Zheng, S. R. Whiteley, H. Kim, J. Kim, X. Meng, and T. Ortlepp, "64-kb Hybrid Josephson-CMOS 4 Kelvin RAM With 400 ps Access Time and 12 mW Read Power," *IEEE Transactions on Applied Superconductivity*, vol. 23, no. 3, p. 1700504, June 2013.

128. H.-S. P. Wong and S. Salahuddin, "Memory Leads the Way to Better Computing," *Nature Nanotechnology*, vol. 10, no. 3, pp. 191–194, March 2015.

129. L. Ye, D. B. Gopman, L. Rehm, D. Backes, G. Wolf, T. Ohki, A. F. Kirichenko, I. V. Vernik, O. A. Mukhanov, and A. D. Kent, "Spin-Transfer Switching of Orthogonal Spin-Valve Devices at Cryogenic Temperatures," *Journal of Applied Physics*, vol. 115, no. 17, p. 17C725, February 2014.

130. L. Liu, C.-F. Pai, Y. Li, H. W. Tseng, D. C. Ralph, and R. A. Buhrman, "Spin-Torque Switching with the Giant Spin Hall Effect of Tantalum," *Science*, vol. 336, no. 6081, pp. 555–558, May 2012.

131. M.-H. Nguyen *et al.*, "Cryogenic Memory Architecture Integrating Spin Hall Effect based Magnetic Memory and Superconductive Cryotron Devices," *Scientific Reports*, vol. 10, no. 1, p. 248, January 2020.

132. O. A. Mukhanov *et al.*, "Superconductor Digital-RF Receiver Systems," *IEICE Transactions on Electronics*, vol. 91, no. 3, pp. 306–317, March 2008.

133. A. Fujimaki, M. Tanaka, T. Yamada, Y. Yamanashi, H. Park, and N. Yoshikawa, "Bit-Serial Single Flux Quantum Microprocessor CORE," *IEICE Transactions on Electronics*, vol. 91, no. 3, pp. 342–349, March 2008.

134. Y. Ando, R. Sato, M. Tanaka, K. Takagi, N. Takagi, and A. Fujimaki, "Design and Demonstration of an 8-bit Bit-Serial RSFQ Microprocessor: CORE e4," *IEEE Transactions on Applied Superconductivity*, vol. 26, no. 5, pp. 1–5, August 2016.

135. A. F. Kirichenko, I. V. Vernik, M. Y. Kamkar, J. Walter, M. Miller, L. R. Albu, and O. A. Mukhanov, "ERSFQ 8-Bit Parallel Arithmetic Logic Unit," *IEEE Transactions on Applied Superconductivity*, vol. 29, no. 5, pp. 1–7, August 2019.

136. V. K. Semenov, Y. A. Polyakov, and S. K. Tolpygo, "AC-Biased Shift Registers as Fabrication Process Benchmark Circuits and Flux Trapping Diagnostic Tool," *IEEE Transactions on Applied Superconductivity*, vol. 27, no. 4, June 2017.

137. T. Jabbari, G. Krylov, S. Whiteley, J. Kawa, and E. G. Friedman, "Global Signaling for Large Scale RSFQ Circuits," *Proceedings of the Government Microcircuit Applications & Critical Technology Conference*, March 2019.

138. B. Dimov, V. Todorov, V. Mladenov, and F. H. Uhlmann, "Optimal Signal Propagation Speed of a Josephson Transmission Line," *Superconductor Science and Technology*, vol. 17, no. 6, pp. 819–822, April 2004.

139. S. K. Tolpygo, V. Bolkhovsky, T. J. Weir, A. Wynn, D. E. Oates, L. M. Johnson, and M. A. Gouker, "Advanced Fabrication Processes for Superconducting Very Large-Scale Integrated Circuits," *IEEE Transactions on Applied Superconductivity*, vol. 26, no. 3, pp. 1–10, April 2016.

140. A. V. Rylyakov, *Ultra-Low-Power RSFQ Devices and Digital Autocorrelation of Broadband Signals*, Ph.D. Dissertation, State University of New York at Stony Brook, New York, 1997.

141. R. L. Kautz, "Picosecond Pulses on Superconducting Striplines," *Journal of Applied Physics*, vol. 49, no. 1, pp. 308–314, January 1978.

142. T. Y. Hsiang, J. F. Whitaker, R. Sobolewski, D. R. Dykaar, and G. A. Mourou, "Propagation Characteristics of Picosecond Electrical Transients on Coplanar Striplines," *Applied Physics Letters*, vol. 51, no. 19, pp. 1551–1553, November 1987.

143. Y. Hashimoto, S. Yorozu, Y. Kameda, A. Fujimaki, H. Terai, and N. Yoshikawa, "Design and Investigation of Gate-to-Gate Passive Interconnections for SFQ Logic Circuits," *IEEE Transactions on Applied Superconductivity*, vol. 15, no. 3, pp. 3814–3820, September 2005.

144. S. V. Polonsky, V. K. Semenov, and D. F. Schneider, "Transmission of Single-Flux-Quantum Pulses along Superconducting Microstrip Lines," *IEEE Transactions on Applied Superconductivity*, vol. 3, no. 1, pp. 2598–2600, March 1993.

145. L. Schindler, P. l. Roux, and C. J. Fourie, "Impedance Matching of Passive Transmission Line Receivers to Improve Reflections Between RSFQ Logic Cells," *IEEE Transactions on Applied Superconductivity*, vol. 30, no. 2, pp. 1–7, March 2020.

146. T. Jabbari, G. Krylov, S. Whiteley, J. Kawa, and E. G. Friedman, "Repeater Insertion in SFQ Interconnect," *IEEE Transactions on Applied Superconductivity*, vol. 30, no. 8, pp. 1–8, Art no. 5 400 508, December 2020.

147. H. Suzuki, S. Nagasawa, K. Miyahara, and Y. Enomoto, "Characteristics of Driver and Receiver Circuits with a Passive Transmission Line in RSFQ Circuits," *IEEE Transactions on Applied Superconductivity*, vol. 10, no. 3, pp. 1637–1641, September 2000.

148. S. Razmkhah and A. Bozbey, "Design of the Passive Transmission Lines for Different Stripline Widths and Impedances," *IEEE Transactions on Applied Superconductivity*, vol. 26, no. 8, pp. 1–6, December 2016.

149. T. V. Filippov and V. K. Kornev, "Sensitivity of the Balanced Josephson-Junction Comparator," *IEEE Transactions on Magnetics*, vol. 27, no. 2, pp. 2452–2455, March 1991.

150. T. V. Filippov, A. Sahu, M. E. Çelik, D. E. Kirichenko, and D. Gupta, "The Josephson Balanced Comparator and its Gray Zone Measurements," *Proceedings of the IEEE International Superconductive Electronics Conference*, pp. 1–3, July 2019.

151. T. Jabbari, G. Krylov, J. Kawa, and E. G. Friedman, "Splitter Trees in Single Flux Quantum Circuits," *IEEE Transactions on Applied Superconductivity*, vol. 31, no. 5, pp. 1–6, Art no. 1 302 606, August 2021.

152. T. Jabbari and E. G. Friedman, "Global Interconnects in VLSI Complexity Single Flux Quantum Systems," *Proceedings of the Workshop on System-Level Interconnect: Problems and Pathfinding Workshop*, November 2020.

153. H. Kumar, T. Jabbari, G. Krylov, K. Basu, E. G. Friedman, and R. Karri, "Toward Increasing the Difficulty of Reverse Engineering of RSFQ Circuits," *IEEE Transactions on Applied Superconductivity*, vol. 30, no. 3, pp. 1–13, Art no. 1 700 213, April 2020.

154. T. Jabbari, G. Krylov, and E. G. Friedman, "Logic Locking in Single Flux Quantum Circuits," *IEEE Transactions on Applied Superconductivity*, vol. 31, no. 5, pp. 1–5, Art no. 1 301 605, August 2021.

155. D. Muller and W. Bartky, "A Theory of Asynchronous Circuits," *Proceedings of the International Symposium on the Theory of Switching*, pp. 204–243, April 1959.

156. D. E. Kirichenko, S. Sarwana, and A. F. Kirichenko, "Zero Static Power Dissipation Biasing of RSFQ Circuits," *IEEE Transactions on Applied Superconductivity*, vol. 21, no. 3, pp. 776–779, January 2011.

157. N. Yoshikawa and Y. Kato, "Reduction of Power Consumption of RSFQ Circuits by Inductance-Load Biasing," *Superconductor Science and Technology*, vol. 12, no. 11, pp. 918–920, November 1999.

158. S. Polonsky, "Delay Insensitive RSFQ Circuits with Zero Static Power Dissipation," *IEEE Transactions on Applied Superconductivity*, vol. 9, no. 2, pp. 3535–3538, June 1999.

159. P. Patra, S. Polonsky, and D. S. Fussell, "Delay Insensitive Logic for RSFQ Superconductor Technology," *Proceedings of the IEEE International Symposium on Advanced Research in Asynchronous Circuits and Systems*, pp. 42–53, April 1997.

160. S. S. Meher, C. Kanungo, A. Shukla, and A. Inamdar, "Parametric Approach for Routing Power Nets and Passive Transmission Lines as Part of Digital Cells," *IEEE Transactions on Applied Superconductivity*, vol. 29, no. 5, pp. 1–7, August 2019.

161. C. Shawawreh, D. Amparo, J. Ren, M. Miller, M. Y. Kamkar, A. Sahu, A. Inamdar, A. F. Kirichenko, O. A. Mukhanov, and I. V. Vernik, "Effects of Adaptive DC Biasing on Operational Margins in ERSFQ Circuits," *IEEE Transactions on Applied Superconductivity*, vol. 27, no. 4, pp. 1–6, June 2017.

162. I. V. Vernik, A. F. Kirichenko, O. A. Mukhanov, and T. A. Ohki, "Energy-Efficient and Compact ERSFQ Decoder for Cryogenic RAM," *IEEE Transactions on Applied Superconductivity*, vol. 27, no. 4, pp. 1–5, June 2017.

163. N. K. Katam, O. Mukhanov, and M. Pedram, "Simulation Analysis and Energy-Saving Techniques for ERSFQ Circuits," *IEEE Transactions on Applied Superconductivity*, vol. 29, no. 5, pp. 1–7, August 2019.

164. M. B. Ketchen, J. Timmerwilke, G. W. Gibson, and M. Bhushan, "ERSFQ Power Delivery: A Self-Consistent Model/Hardware Case Study," *IEEE Transactions on Applied Superconductivity*, vol. 29, no. 7, pp. 1–11, October 2019.

165. G. Li, J. Ren, Y. Wu, L. Ying, M. Niu, L. Chen, and Z. Wang, "Research on the Bias Network of Energy-Efficient Single Flux Quantum Circuits," *IEEE Transactions on Applied Superconductivity*, vol. 29, no. 5, pp. 1–5, August 2019.

166. G. Krylov and E. G. Friedman, "Bias Networks for High Complexity Energy Efficient Single Flux Quantum Circuits," *Proceedings of the Government Microcircuit Applications & Critical Technology Conference*, March 2020.

167. M. H. Volkmann, A. Sahu, C. J. Fourie, and O. A. Mukhanov, "Implementation of Energy Efficient Single Flux Quantum Digital Circuits with Sub-aJ/bit Operation," *Superconductor Science and Technology*, vol. 26, no. 1, p. 015002, November 2012.

168. Q. P. Herr and P. Bunyk, "Implementation and Application of First-In First-Out Buffers," *IEEE Transactions on Applied Superconductivity*, vol. 13, no. 2, pp. 563–566, June 2003.

169. T. Jabbari, E. G. Friedma, and J. Kawa, "H-Tree Clock Synthesis in RSFQ Circuits," *Proceedings of the IEEE Baltic Electronics Conference*, pp. 1–5, October 2020.

170. E. G. Friedman, *High Performance Clock Distribution Networks*, Springer, 1997.

171. K. Gaj, E. G. Friedman, and M. J. Feldman, "Timing of Large RSFQ Digital Circuits," *Proceedings of the IEEE International Superconductive Electronics Conference*, pp. 299–301, June 1997.

172. Y. Kameda, S. Yorozu, and Y. Hashimoto, "A New Design Methodology for Single-Flux-Quantum (SFQ) Logic Circuits using Passive-Transmission-Line (PTL) Wiring," *IEEE Transactions on Applied Superconductivity*, vol. 17, no. 2, pp. 508–511, June 2007.

173. S. N. Shahsavani and M. Pedram, "A Minimum-Skew Clock Tree Synthesis Algorithm for Single Flux Quantum Logic Circuits," *IEEE Transactions on Applied Superconductivity*, vol. 29, no. 8, pp. 1–13, September 2019.

174. J. L. Neves and E. G. Friedman, "Automated Synthesis of Skew-Based Clock Distribution Networks," *VLSI Design*, vol. 7, no. 1, pp. 31–57, 1998.

175. Q. P. Herr, N. Vukovic, C. A. Mancini, K. Gaj, Qing Ke, V. Adler, E. G. Friedman, A. Krasniewski, M. F. Bocko, and M. J. Feldman, "Design and Low Speed Testing of a Four-Bit RSFQ Multiplier-Accumulator," *IEEE Transactions on Applied Superconductivity*, vol. 7, no. 2, pp. 3168–3171, June 1997.

176. K. Gaj, E. G. Friedman, M. J. Feldman, and A. Krasniewski, "A Clock Distribution Scheme for Large RSFQ Circuits," *IEEE Transactions on Applied Superconductivity*, vol. 5, no. 2, pp. 3320–3324, June 1995.

177. P. Bunyk and P. Litskevitch, "Case Study in RSFQ Design: Fast Pipelined Parallel Adder," *IEEE Transactions on Applied Superconductivity*, vol. 9, no. 2, pp. 3714–3720, June 1999.

178. S. N. Shahsavani, T. Lin, A. Shafaei, C. J. Fourie, and M. Pedram, "An Integrated Row-Based Cell Placement and Interconnect Synthesis Tool for Large SFQ Logic Circuits," *IEEE Transactions on Applied Superconductivity*, vol. 27, no. 4, pp. 1–8, March 2017.

179. S. N. Shahsavani, A. Shafaei, and M. Pedram, "A Placement Algorithm for Superconducting Logic Circuits Based on Cell Grouping and Super-Cell Placement," *Proceedings of the ACM/IEEE Design, Automation & Test in Europe Conference*, pp. 1465–1468, March 2018.

180. E. G. Friedman, "Clock Distribution Design in VLSI Circuits – an Overview," *Proceedings of the IEEE International Symposium on Circuits and Systems*, vol. 3, pp. 1475–1478, May 1993.

181. E. G. Friedman, "Clock Distribution Networks in Synchronous Digital Integrated Circuits," *Proceedings of the IEEE*, vol. 89, no. 5, pp. 665–692, May 2001.

182. O. A. Mukhanov, "Rapid Single Flux Quantum (RSFQ) Shift Register Family," *IEEE Transactions on Applied Superconductivity*, vol. 3, no. 1, pp. 2578–2581, March 1993.

183. C. A. Mancini, N. Vukovic, A. M. Herr, K. Gaj, M. F. Bocko, and M. J. Feldman, "RSFQ Circular Shift Registers," *IEEE Transactions on Applied Superconductivity*, vol. 7, no. 2, pp. 2832–2835, June 1997.

184. R. N. Tadros and P. A. Beerel, "A Robust and Tree-Free Hybrid Clocking Technique for RSFQ Circuits – CSR Application," *Proceedings of the IEEE International Superconductive Electronics Conference*, pp. 1–4, March 2017.

185. R. N. Tadros and P. A. Beerel, "A Robust and Self-Adaptive Clocking Technique for SFQ Circuits," *IEEE Transactions on Applied Superconductivity*, vol. 28, no. 7, pp. 1–11, October 2018.

186. E. G. Friedman (Ed.), *Clock Distribution Networks in VLSI Circuits and Systems*, IEEE Press, 1995.

187. M. Dorojevets, P. Bunyk, and D. Zinoviev, "FLUX Chip: Design of a 20-GHz 16-bit Ultra-pipelined RSFQ Processor Prototype based on 1.75-μm LTS Technology," *IEEE Transactions on Applied Superconductivity*, vol. 11, no. 1, pp. 326–332, March 2001.

188. M. Dorojevets and P. Bunyk, "Architectural and Implementation Challenges in Designing High-Performance RSFQ Processors: a FLUX-1 Microprocessor and Beyond," *IEEE Transactions on Applied Superconductivity*, vol. 13, no. 2, pp. 446–449, June 2003.

189. S. V. Rylov, "Clockless Dynamic SFQ and Gate With High Input Skew Tolerance," *IEEE Transactions on Applied Superconductivity*, vol. 29, no. 5, pp. 1–5, August 2019.

190. L. C. Müller, H. R. Gerber, and C. J. Fourie, "Review and Comparison of RSFQ Asynchronous Methodologies," *Journal of Physics: Conference Series*, vol. 97, p. 012109, February 2008.

191. I. E. Sutherland, "Micropipelines," *Communications of the ACM*, vol. 32, no. 6, p. 720–738, June 1989.

192. O. A. Mukhanov, S. V. Rylov, V. K. Semonov, and S. V. Vyshenskii, "RSFQ Logic Arithmetic," *IEEE Transactions on Magnetics*, vol. 25, no. 2, pp. 857–860, March 1989.

193. T. V. Filippov, A. Sahu, A. F. Kirichenko, I. V. Vernik, M. Dorojevets, C. L. Ayala, and O. A. Mukhanov, "20 GHz Operation of an Asynchronous Wave-Pipelined RSFQ Arithmetic-Logic Unit," *Physics Procedia*, vol. 36, pp. 59–65, May 2012.

194. M. Dorojevets, C. Ayala, and A. Kasperek, "Development and Evaluation of Design Techniques for High-Performance Wave-Pipelined Wide Datapath RSFQ Processors," *Proceedings of the IEEE International Superconductive Electronics Conference*, pp. 1–2, June 2009.

195. T. Filippov, M. Dorojevets, A. Sahu, A. Kirichenko, C. Ayala, and O. Mukhanov, "8-Bit Asynchronous Wave-Pipelined RSFQ Arithmetic-Logic Unit," *IEEE Transactions on Applied Superconductivity*, vol. 21, no. 3, pp. 847–851, February 2011.

196. M. Dorojevets, C. L. Ayala, N. Yoshikawa, and A. Fujimaki, "16-Bit Wave-Pipelined Sparse-Tree RSFQ Adder," *IEEE Transactions on Applied Superconductivity*, vol. 23, no. 3, p. 1700605, June 2013.

197. M. Maezawa, I. Kurosawa, M. Aoyagi, H. Nakagawa, Y. Kameda, and T. Nanya, "Rapid Single-Flux-Quantum Dual-Rail Logic for Asynchronous Circuits," *IEEE Transactions on Applied Superconductivity*, vol. 7, no. 2, pp. 2705–2708, June 1997.

198. I. Kurosawa, H. Nakagawa, M. Aoyagi, M. Maezawa, Y. Kameda, and T. Nanya, "A Basic Circuit for Asynchronous Superconductive Logic Using RSFQ Gates," *Superconductor Science and Technology*, vol. 9, no. 4A, pp. A46–A49, April 1996.

199. T. Hosoki, H. Kodaka, M. Kitagawa, and Y. Okabe, "Design and Experimentation of BSFQ Logic Devices," *Superconductor Science and Technology*, vol. 12, no. 11, pp. 773–775, November 1999.

200. H. R. Gerber, C. J. Fourie, and W. J. Perold, "RSFQ-Asynchronous Timing (RSFQ-AT): a New Design Methodology for Implementation in CAD Automation," *IEEE Transactions on Applied Superconductivity*, vol. 15, no. 2, pp. 272–275, June 2005.

201. Z. J. Deng, N. Yoshikawa, S. R. Whiteley, and T. Van Duzer, "Data-Driven Self-Timed RSFQ Digital Integrated Circuit and System," *IEEE Transactions on Applied Superconductivity*, vol. 7, no. 2, pp. 3634–3637, June 1997.

202. N. Yoshikawa, F. Matsuzaki, N. Nakajima, K. Fujiwara, K. Yoda, and K. Kawasaki, "Design and Component Test of a Tiny Processor Based on the SFQ Technology," *IEEE Transactions on Applied Superconductivity*, vol. 13, no. 2, pp. 441–445, July 2003.

203. Y. Nobumori, T. Nishigai, K. Nakamiya, N. Yoshikawa, A. Fujimaki, H. Terai, and S. Yorozu, "Design and Implementation of a Fully Asynchronous SFQ Microprocessor: SCRAM2," *IEEE Transactions on Applied Superconductivity*, vol. 17, no. 2, pp. 478–481, June 2007.

204. R. M. Keller, "Towards a Theory of Universal Speed-Independent Modules," *IEEE Transactions on Computers*, vol. C-23, no. 1, pp. 21–33, January 1974.

205. N. Tsuji, Y. Yamanashi, N. Takeuchi, C. Ayala, and N. Yoshikawa, "Design and Implementation of Scalable Register Files Using Adiabatic Quantum Flux Parametron Logic," *Proceedings of the IEEE International Superconductive Electronics Conference*, pp. 1–3, June 2017.

206. N. Takeuchi, M. Nozoe, Y. He, and N. Yoshikawa, "Low-Latency Adiabatic Superconductor Logic using Delay-Line Clocking," *Applied Physics Letters*, vol. 115, no. 7, p. 072601, August 2019.

207. Y. He, N. Takeuchi, and N. Yoshikawa, "Low-Latency Power-Dividing Clocking Scheme for Adiabatic Quantum-Flux-Parametron Logic," *Applied Physics Letters*, vol. 116, no. 18, p. 182602, May 2020.

208. N. Takeuchi, C. L. Ayala, O. Chen, and N. Yoshikawa, "A Feedback-Friendly Large-Scale Clocking Scheme for Adiabatic Quantum-Flux-Parametron Logic Datapaths," *IEEE Transactions on Applied Superconductivity*, vol. 29, no. 5, pp. 1–5, August 2019.

209. S. K. Tolpygo, V. Bolkhovsky, R. Rastogi, S. Zarr, A. L. Day, E. Golden, T. J. Weir, A. Wynn, and L. M. Johnson, "Advanced Fabrication Processes for Superconductor Electronics: Current Status and New Developments," *IEEE Transactions on Applied Superconductivity*, vol. 29, no. 5, pp. 1–13, August 2019.

210. T. Ando, S. Nagasawa, N. Takeuchi, N. Tsuji, F. China, M. Hidaka, Y. Yamanashi, and N. Yoshikawa, "Three-Dimensional Adiabatic Quantum-Flux-Parametron Fabricated using a Double-Active-Layered Niobium Process," *Superconductor Science and Technology*, vol. 30, no. 7, p. 075003, June 2017.

211. D. T. Yohannes, R. T. Hunt, J. A. Vivalda, D. Amparo, A. Cohen, I. V. Vernik, and A. F. Kirichenko, "Planarized, Extendible, Multilayer Fabrication Process for Superconducting Electronics," *IEEE Transactions on Applied Superconductivity*, vol. 25, no. 3, pp. 1–5, June 2015.

212. D. Yohannes, S. Sarwana, S. K. Tolpygo, A. Sahu, Y. A. Polyakov, and V. K. Semenov, "Characterization of HYPRES' 4.5 kA/cm^2 & 8 kA/cm^2 $Nb/AlO_x/Nb$ Fabrication Processes," *IEEE Transactions on Applied Superconductivity*, vol. 15, no. 2, pp. 90–93, June 2005.

213. F. Bedard, N. Welker, G.R. Gotter, M.A. Escavage, and J.T. Pinkston, *Superconducting Technology Assessment*, National Security Agency, Office of Corporate Assessments, Fort Meade, Maryland, USA, 2005.

214. J. M. Murduck, "Fabrication of Superconducting Devices and Circuits," *Frontiers of Thin Film Technology*, Elsevier, 2001.

215. Y. Tarutani, M. Hirano, and U. Kawabe, "Niobium-Based Integrated Circuit Technologies," *Proceedings of the IEEE*, vol. 77, no. 8, pp. 1164–1176, August 1989.

216. E. L. Wolf, "Introduction to Refractory Josephson Junctions," *Josephson Junctions: History, Devices, and Applications*, E. L. Wolf, G. B. Arnold, M. A. Gurvitch and J. F. Zasadzinski (Eds.), Pan Stanford Publishing Pte. Ltd., Chapter 2, pp. 17 - 46, 2017.

217. M. A. Gurvitch, "The Trace That Launched a Thousand Chips: Development of Nb/Al–Oxide–Nb Technology," *Josephson Junctions: History, Devices, and Applications*, E. L. Wolf, G. B. Arnold, M. A. Gurvitch and J. F. Zasadzinski (Eds.), Pan Stanford Publishing Pte. Ltd., Chapter 5, pp. 83 - 146, 2017.

218. A. L. Robinson, "New Superconductors for a Supercomputer," *Science*, vol. 215, no. 4528, pp. 40–43, January 1982.

219. I. Ames, "An Overview of Materials and Process Aspects of Josephson Integrated Circuit Fabrication," *IBM Journal of Research and Development*, vol. 24, no. 2, pp. 188–194, March 1980.

220. I. Giaever, "Energy Gap in Superconductors Measured by Electron Tunneling," *Physical Review Letters*, vol. 5, pp. 147–148, August 1960.

221. D. Shen, R. Zhu, W. Xu, J. Chang, Z. Ji, G. Sun, C. Cao, and J. Chen, "Character and Fabrication of Al/Al_2O_3/Al Tunnel Junctions for Qubit Application," *Chinese Science Bulletin*, vol. 57, no. 4, pp. 409–412, February 2012.

222. D.-R. W. Yost *et al.*, "Solid-state Qubits Integrated with Superconducting Through-Silicon Vias," *npj Quantum Information*, vol. 6, no. 1, pp. 1–7, July 2020.

223. M. Gurvitch, M. A. Washington, and H. A. Huggins, "High Quality Refractory Josephson Tunnel Junctions utilizing Thin Aluminum Layers," *Applied Physics Letters*, vol. 42, no. 5, pp. 472–474, March 1983.

224. Y. Uzawa, S. Saito, W. Qiu, K. Makise, T. Kojima, and Z. Wang, "Optical and Tunneling Studies of Energy Gap in Superconducting Niobium Nitride Films," *Journal of Low Temperature Physics*, vol. 199, pp. 143–148, January 2020.

225. M. M. T. M. Dierichs, B. J. Feenstra, A. Skalare, C. E. Honingh, J. Mees, H. Stadt, and T. d. Graauw, "Evaluation of Niobium Transmission Lines up to the Superconducting Gap Frequency," *Applied Physics Letters*, vol. 63, no. 2, pp. 249–251, June 1993.

226. J.-C. Villegier, "Refractory Niobium Nitride NbN Josephson Junctions and Applications," *Josephson Junctions: History, Devices, and Applications*, E. L. Wolf, G. B. Arnold, M. A. Gurvitch and J. F. Zasadzinski (Eds.), Pan Stanford Publishing Pte. Ltd., Chapter 6, pp. 147 - 183, 2017.

227. M. Radparvar, "Superconducting Niobium and Niobium Nitride Processes for Medium-Scale Integration Applications," *Cryogenics*, vol. 35, pp. 535–540, August 1995.

228. L. A. Abelson and G. L. Kerber, "Superconductor Integrated Circuit Fabrication Technology," *Proceedings of the IEEE*, vol. 92, no. 10, pp. 1517–1533, October 2004.

229. D. C. Rorer, D. G. Onn, and H. Meyer, "Thermodynamic Properties of Molybdenum in its Superconducting and Normal State," *Physical Review*, vol. 138, pp. A1661–A1668, June 1965.

230. S. K. Tolpygo, V. Bolkhovsky, T. J. Weir, L. M. Johnson, M. A. Gouker, and W. D. Oliver, "Fabrication Process and Properties of Fully-Planarized Deep-Submicron Nb/Al–AlO_x/Nb Josephson Junctions for VLSI Circuits," *IEEE Transactions on Applied Superconductivity*, vol. 25, no. 3, pp. 1–12, June 2015.

231. W. L. McMillan, "Tunneling Model of the Superconducting Proximity Effect," *Physical Review*, vol. 175, pp. 537–542, November 1968.

232. S. K. Tolpygo and D. Amparo, "Electrical Stress Effect on Josephson Tunneling through Ultrathin AlOx Barrier in Nb/Al/AlOx/Nb Junctions," *Journal of Applied Physics*, vol. 104, no. 6, p. 063904, September 2008.

233. R. E. Miller, W. H. Mallison, A. W. Kleinsasser, K. A. Delin, and E. M. Macedo, "Niobium Trilayer Josephson Tunnel Junctions with Ultrahigh Critical Current Densities," *Applied Physics Letters*, vol. 63, no. 10, pp. 1423–1425, September 1993.

234. V. F. Pavlidis, I. Savidis, and E. G. Friedman, *Three-Dimensional Integrated Circuit Design, Second Edition*, Morgan Kaufmann, 2017.

235. H. Jun, J. Cho, K. Lee, H. Son, K. Kim, H. Jin, and K. Kim, "HBM (High Bandwidth Memory) DRAM Technology and Architecture," *Proceedings of the IEEE International Memory Workshop*, May 2017.

236. C. Monzio Compagnoni, A. Goda, A. S. Spinelli, P. Feeley, A. L. Lacaita, and A. Visconti, "Reviewing the Evolution of the NAND Flash Technology," *Proceedings of the IEEE*, vol. 105, no. 9, pp. 1609–1633, September 2017.

237. B. Vaisband, *3-D ICs as a Platform for Heterogeneous Systems Integration*, Ph.D. Dissertation, University of Rochester, Rochester, New York, 2017.

238. S. K. Tolpygo, V. Bolkhovsky, R. Rastogi, S. Zarr, A. L. Day, E. Golden, T. J. Weir, A. Wynn, and L. M. Johnson, "Planarized Fabrication Process With Two Layers of SIS Josephson Junctions and Integration of SIS and SFS π-Junctions," *IEEE Transactions on Applied Superconductivity*, vol. 29, no. 5, pp. 1–8, August 2019.

239. K. Gaj, Q. P. Herr, V. Adler, D. K. Brock, E. G. Friedman, and M. J. Feldman, "Toward a Systematic Design Methodology for Large Multigigahertz Rapid Single Flux Quantum Circuits," *IEEE Transactions on Applied Superconductivity*, vol. 9, no. 3, pp. 4591–4606, September 1999.

240. A. B. Kahng, J. Lienig, I. L. Markov, and J. Hu, *VLSI Physical Design: From Graph Partitioning to Timing Closure*, Springer Netherlands, 2011.
241. Stony Brook University. SUNY RSFQ Cell Library. [Online]. Available: http://www.physics.sunysb.edu/physics/rsfq/lib/contents.html
242. Technische Universität Ilmenau. RSFQ - Cell Library. [Online]. Available: https://www.tu-ilmenau.de/en/advanced-electromagnetics-group/research/superconductive-high-speed-electronics/rsfq-cell/
243. S. Anders *et al.*, "European Roadmap on Superconductive Electronics – Status and Perspectives," *Physica C: Superconductivity*, vol. 470, no. 23, pp. 2079–2126, December 2010.
244. S. Yorozu, Y. Kameda, H. Terai, A. Fujimaki, T. Yamada, and S. Tahara, "A Single Flux Quantum Standard Logic Cell Library," *Physica C: Superconductivity*, vol. 378-381, pp. 1471–1474, October 2002.
245. S. Tahara, H. Numata, S. Yorozu, Y. Hashimoto, and S. Nagasawa, "Superconducting Technology for Digital Applications using Niobium Josephson Junctions," *IEICE Transactions on Electronics*, vol. 83, no. 1, pp. 60–68, January 2000.
246. M. Maezawa, M. Ochiai, H. Kimura, F. Hirayama, and M. Suzuki, "Design and Operation of RSFQ Cell Library Fabricated by Using a 10-kA/cm^2 Nb Technology," *IEEE Transactions on Applied Superconductivity*, vol. 17, no. 2, pp. 500–504, June 2007.
247. M. Maezawa, F. Hirayama, and M. Suzuki, "Design and Fabrication of RSFQ Cell Library for Middle-Scale Applications," *Physica C: Superconductivity*, vol. 412-414, pp. 1591–1596, October 2004.
248. A. Inamdar, D. Amparo, B. Sahoo, J. Ren, and A. Sahu, "RSFQ/ERSFQ Cell Library With Improved Circuit Optimization, Timing Verification, and Test Characterization," *IEEE Transactions on Applied Superconductivity*, vol. 27, no. 4, pp. 1–9, June 2017.
249. D. Amparo, M. Eren Çelik, S. Nath, J. P. Cerqueira, and A. Inamdar, "Timing Characterization for RSFQ Cell Library," *IEEE Transactions on Applied Superconductivity*, vol. 29, no. 5, pp. 1–9, August 2019.
250. C. L. Ayala, R. Saito, T. Tanaka, O. Chen, N. Takeuchi, Y. He, and N. Yoshikawa, "A Semi-Custom Design Methodology and Environment for Implementing Superconductor Adiabatic Quantum-Flux-Parametron Microprocessors," *Superconductor Science and Technology*, vol. 33, no. 5, p. 054006, March 2020.
251. Y. He, C. L. Ayala, N. Takeuchi, T. Yamae, Y. Hironaka, A. Sahu, V. Gupta, A. Talalaevskii, D. Gupta, and N. Yoshikawa, "A Compact AQFP Logic Cell Design Using an 8-Metal Layer Superconductor Process," *Superconductor Science and Technology*, vol. 33, no. 3, p. 035010, February 2020.
252. C. L. Ayala, O. Chen, and N. Yoshikawa, "AQFPTX: Adiabatic Quantum-Flux-Parametron Timing eXtraction Tool," *Proceedings of the IEEE International Superconductive Electronics Conference*, pp. 1–3, August 2019.
253. K. Gaj, C.-H. Cheah, E. G. Friedman, and M. J. Feldman, "Functional Modeling of RSFQ Circuits using Verilog HDL," *IEEE Transactions on Applied Superconductivity*, vol. 7, no. 2, pp. 3151–3154, June 1997.
254. A. Krasniewski, "Logic Simulation of RSFQ Circuits," *IEEE Transactions on Applied Superconductivity*, vol. 3, no. 1, pp. 33–38, March 1993.
255. S. V. Polonsky, V. K. Semenov, and P. N. Shevchenko, "PSCAN: Personal Superconductor Circuit Analyser," *Superconductor Science and Technology*, vol. 4, no. 11, pp. 667–670, November 1991.
256. P. Bunyk, A. Y. Kidiyarova-Shevchenko, and P. Litskevitch, "RSFQ Microprocessor: New Design Approaches," *IEEE Transactions on Applied Superconductivity*, vol. 7, no. 2, pp. 2697–2704, June 1997.
257. H. Toepfer, T. Harnisch, J. Kunert, S. Lange, and H. F. Uhlmann, "Formal Description of the Functional Behavior of RSFQ Logic Circuits for Design and Optimization Purposes," *IEEE Transactions on Applied Superconductivity*, vol. 7, no. 2, pp. 3630–3633, June 1997.

258. F. Matsuzaki, N. Yoshikawa, M. Tanaka, A. Fujimaki, and Y. Takai, "A Behavioral-Level HDL Description of SFQ Logic Circuits for Quantitative Performance Analysis of Large-Scale SFQ Digital Systems," *Physica C: Superconductivity*, vol. 392-396, pp. 1495–1500, October 2003.

259. S. Intiso, I. Kataeva, E. Tolkacheva, Henrik Engseth, K. Platov, and A. Kidiyarova-Shevchenko, "Time-Delay Optimization of RSFQ Cells," *IEEE Transactions on Applied Superconductivity*, vol. 15, no. 2, pp. 328–331, June 2005.

260. A. K. Kasperek, *32-bit Superconductor Integer and Floating-Point Multipliers*, Ph.D. Dissertation, Stony Brook University, Stony Brook, New York, 2012.

261. L. C. Müller and C. J. Fourie, "Automated State Machine and Timing Characteristic Extraction for RSFQ Circuits," *IEEE Transactions on Applied Superconductivity*, vol. 24, no. 1, pp. 3–12, February 2014.

262. C. J. Fourie, "Extraction of DC-Biased SFQ Circuit Verilog Models," *IEEE Transactions on Applied Superconductivity*, vol. 28, no. 6, pp. 1–11, September 2018.

263. Q. Xu, C. L. Ayala, N. Takeuchi, Y. Yamanashi, and N. Yoshikawa, "HDL-Based Modeling Approach for Digital Simulation of Adiabatic Quantum Flux Parametron Logic," *IEEE Transactions on Applied Superconductivity*, vol. 26, no. 8, pp. 1–5, December 2016.

264. R. N. Tadros, A. Fayyazi, M. Pedram, and P. A. Beerel, "SystemVerilog Modeling of SFQ and AQFP Circuits," *IEEE Transactions on Applied Superconductivity*, vol. 30, no. 2, pp. 1–13, March 2020.

265. L. W. Nagel, and D. O. Pederson, "SPICE (Simulation Program with Integrated Circuit Emphasis)," EECS Department, University of California, Berkeley, Technical Report UCB/ERL M382, April 1973.

266. S. R. Whiteley, "Josephson Junctions in SPICE3," *IEEE Transactions on Magnetics*, vol. 27, no. 2, pp. 2902–2905, March 1991.

267. E. S. Fang and T. V. Duzer, "A Josephson Integrated Circuit Simulator (JSIM) for Superconductive Electronics Application," *Proceedings of the IEEE International Superconductive Electronics Conference*, pp. 407–410, June 1989.

268. S. R. Whiteley. WRspice Reference Manual. Whiteley Research Inc. [Online]. Available: http://www.wrcad.com/manual/wrsmanual.pdf

269. K. Gaj, Q. P. Herr, V. Adler, A. Krasniewski, E. G. Friedman, and M. J. Feldman, "Tools for the Computer-Aided Design of Multigigahertz Superconducting Digital Circuits," *IEEE Transactions on Applied Superconductivity*, vol. 9, no. 1, pp. 18–38, March 1999.

270. J. A. Delport, K. Jackman, P. l. Roux, and C. J. Fourie, "JoSIM – Superconductor SPICE Simulator," *IEEE Transactions on Applied Superconductivity*, vol. 29, no. 5, pp. 1–5, August 2019.

271. S. Polonsky, P. Shevchenko, A. Kirichenko, D. Zinoviev, and A. Rylyakov, "PSCAN'96: New Software for Simulation and Optimization of Complex RSFQ Circuits," *IEEE Transactions on Applied Superconductivity*, vol. 7, no. 2, pp. 2685–2689, June 1997.

272. P. Shevchenko. PSCAN2 Superconductor Circuit Simulator. [Online]. Available: http://www.pscan2sim.org/documentation.html

273. N. R. Werthamer, "Nonlinear Self-Coupling of Josephson Radiation in Superconducting Tunnel Junctions," *Physical Review*, vol. 147, pp. 255–263, July 1966.

274. A. Odintsov, V. Semenov, and A. Zorin, "Specific Problems of Numerical Analysis of the Josephson Junction Circuits," *IEEE Transactions on Magnetics*, vol. 23, no. 2, pp. 763–766, March 1987.

275. A. De Lustrac, P. Crozat, and R. Adde, "A Picosecond Josephson Junction Model for Circuit Simulation," *Revue de Physique Appliquée*, vol. 21, no. 5, pp. 319–326, May 1986.

276. S. K. Tolpygo, V. Bolkhovsky, T. J. Weir, C. J. Galbraith, L. M. Johnson, M. A. Gouker, and V. K. Semenov, "Inductance of Circuit Structures for MIT LL Superconductor Electronics Fabrication Process With 8 Niobium Layers," *IEEE Transactions on Applied Superconductivity*, vol. 25, no. 3, pp. 1–5, June 2015.

277. Sonnet User's Guide. Sonnet Software Inc. [Online]. Available: https://www.sonnetsoftware.com/support/downloads/manuals/st_users.pdf

278. J. C. Rautio and R. F. Harrington, "An Electromagnetic Time-Harmonic Analysis of Shielded Microstrip Circuits," *IEEE Transactions on Microwave Theory and Techniques*, vol. 35, no. 8, pp. 726–730, August 1987.

279. A. R. Kerr, "Surface Impedance of Superconductors and Normal Conductors in EM Simulators," *National Radio Astronomy Observatory, Electronics Division Internal Report*, no. 302, February 1996.

280. 3D Electromagnetic Field Simulator for RF and Wireless Design. Ansys Inc. [Online]. Available: https://www.ansys.com/products/electronics/ansys-hfss

281. K. U-Yen, K. Rostem, and E. J. Wollack, "Modeling Strategies for Superconducting Microstrip Transmission Line Structures," *IEEE Transactions on Applied Superconductivity*, vol. 28, no. 6, pp. 1–5, September 2018.

282. M. Kamon, M. J. Tsuk, and J. K. White, "FASTHENRY: a Multipole-Accelerated 3-D Inductance Extraction Program," *IEEE Transactions on Microwave Theory and Techniques*, vol. 42, no. 9, pp. 1750–1758, September 1994.

283. I. P. Vaisband, R. Jakushokas, M. Popovich, A. V. Mezhiba, S. Köse, and E. G. Friedman, *On-Chip Power Delivery and Management, Fourth Edition*, Springer, 2016.

284. B. Guan, M. J. Wengler, P. Rott, and M. J. Feldman, "Inductance Estimation for Complicated Superconducting Thin Film Structures with a Finite Segment Method," *IEEE Transactions on Applied Superconductivity*, vol. 7, no. 2, pp. 2776–2779, June 1997.

285. Stephen R. Whiteley. FastHenry 3.0wr. [Online]. Available: http://www.wrcad.com/ftp/pub/readme.fasthenry

286. C. J. Fourie, O. Wetzstein, T. Ortlepp, and J. Kunert, "Three-Dimensional Multi-Terminal Superconductive Integrated Circuit Inductance Extraction," *Superconductor Science and Technology*, vol. 24, no. 12, p. 125015, November 2011.

287. K. Jackman and C. J. Fourie, "Fast Multicore FastHenry and a Tetrahedral Modeling Method for Inductance Extraction of Complex 3D Geometries," *Proceedings of the IEEE International Superconductive Electronics Conference*, pp. 1–3, July 2015.

288. K. Jackman and C. J. Fourie, "Tetrahedral Modeling Method for Inductance Extraction of Complex 3-D Superconducting Structures," *IEEE Transactions on Applied Superconductivity*, vol. 26, no. 3, pp. 1–5, April 2016.

289. M. M. Khapaev, "Inductance Extraction of Multilayer Finite-Thickness Superconductor Circuits," *IEEE Transactions on Microwave Theory and Techniques*, vol. 49, no. 1, pp. 217–220, January 2001.

290. M. M. Khapaev, A. Y. Kidiyarova-Shevchenko, P. Magnelind, and M. Y. Kupriyanov, "3D-MLSI: Software Package for Inductance Calculation in Multilayer Superconducting Integrated Circuits," *IEEE Transactions on Applied Superconductivity*, vol. 11, no. 1, pp. 1090–1093, March 2001.

291. M. M. Khapaev, M. Y. Kupriyanov, E. Goldobin, and M. Siegel, "Current Distribution Simulation for Superconducting Multi-Layered Structures," *Superconductor Science and Technology*, vol. 16, no. 1, pp. 24–27, November 2002.

292. M. M. Khapaev and M. Y. Kupriyanov, "Inductance Extraction of Superconductor Structures with Internal Current Sources," *Superconductor Science and Technology*, vol. 28, no. 5, p. 055013, April 2015.

293. N. Yoshikawa and J. Koshiyama, "Top-Down RSFQ Logic Design Based on a Binary Decision Diagram," *IEEE Transactions on Applied Superconductivity*, vol. 11, no. 1, pp. 1098–1101, March 2001.

294. S. B. Akers, "Binary Decision Diagrams," *IEEE Transactions on Computers*, vol. C-27, no. 6, pp. 509–516, June 1978.

295. ABC: A System for Sequential Synthesis and Verification. Berkeley Logic Synthesis and Verification Group. [Online]. Available: http://www.eecs.berkeley.edu/~alanmi/abc/

296. J. A. Darringer, W. H. Joyner, C. L. Berman, and L. Trevillyan, "Logic Synthesis Through Local Transformations," *IBM Journal of Research and Development*, vol. 25, no. 4, pp. 272–280, July 1981.

297. L. Amarú, P. Gaillardon, and G. De Micheli, "Majority-Inverter Graph: A New Paradigm for Logic Optimization," *IEEE Transactions on Computer-Aided Design of Integrated Circuits and Systems*, vol. 35, no. 5, pp. 806–819, May 2016.

298. K. Inoue, N. Takeuchi, K. Ehara, Y. Yamanashi, and N. Yoshikawa, "Simulation and Experimental Demonstration of Logic Circuits Using an Ultra-Low-Power Adiabatic Quantum-Flux-Parametron," *IEEE Transactions on Applied Superconductivity*, vol. 23, no. 3, p. 1301105, June 2013.

299. L. Amarú, P. Gaillardon, A. Chattopadhyay, and G. De Micheli, "A Sound and Complete Axiomatization of Majority-*n* Logic," *IEEE Transactions on Computers*, vol. 65, no. 9, pp. 2889–2895, September 2016.

300. C. Wolf. Yosys Open Synthesis Suite. [Online]. Available: http://www.clifford.at/yosys/

301. Q. Xu, C. L. Ayala, N. Takeuchi, Y. Murai, Y. Yamanashi, and N. Yoshikawa, "Synthesis Flow for Cell-Based Adiabatic Quantum-Flux-Parametron Structural Circuit Generation With HDL Back-End Verification," *IEEE Transactions on Applied Superconductivity*, vol. 27, no. 4, pp. 1–5, January 2017.

302. M. Pedram and Y. Wang, "Design Automation Methodology and Tools for Superconductive Electronics," *Proceedings of the IEEE/ACM International Conference on Computer-Aided Design*, pp. 1–6, November 2018.

303. N. Katam, A. Shafaei, and M. Pedram, "Design of Complex Rapid Single-Flux-Quantum Cells with Application to Logic Synthesis," *Proceedings of the IEEE International Superconductive Electronics Conference*, pp. 1–3, June 2017.

304. G. Pasandi and M. Pedram, "PBMap: A Path Balancing Technology Mapping Algorithm for Single Flux Quantum Logic Circuits," *IEEE Transactions on Applied Superconductivity*, vol. 29, no. 4, pp. 1–14, November 2019.

305. Xun Liu, M. C. Papaefthymiou, and E. G. Friedman, "Retiming and Clock Scheduling for Digital Circuit Optimization," *IEEE Transactions on Computer-Aided Design of Integrated Circuits and Systems*, vol. 21, no. 2, pp. 184–203, August 2002.

306. T. Soyata, E. G. Friedman, and J. H. Mulligan Jr., "Incorporating Interconnect, Register, and Clock Distribution Delays into the Retiming Process," *IEEE Transactions on Computer-Aided Design of Integrated Circuits and Systems*, vol. 16, no. 1, pp. 105–120, January 1997.

307. N. Kito, K. Takagi, and N. Takagi, "Conversion of a CMOS Logic Circuit Design to an RSFQ Design Considering Latching Function of RSFQ Logic Gates," *IEEE Transactions on Applied Superconductivity*, vol. 25, no. 3, pp. 1–5, June 2015.

308. C. M. Fiduccia and R. M. Mattheyses, "A Linear-Time Heuristic for Improving Network Partitions," *Proceedings of the ACM/IEEE Design Automation Conference*, pp. 175–181, June 1982.

309. S. Whiteley, E. Mlinar, G. Krylov, T. Jabbari, E. G. Friedman, and J. Kawa, "An SFQ Digital Circuit Technology with Fully-Passive Transmission Line Interconnect," *Proceedings of the Applied Superconductivity Conference*, November 2020.

310. M. Tanaka, K. Obata, Y. Ito, S. Takeshima, M. Sato, K. Takagi, N. Takagi, H. Akaike, and A. Fujimaki, "Automated Passive-Transmission-Line Routing Tool for Single-Flux-Quantum Circuits Based on A* Algorithm," *IEICE Transactions on Electronics*, vol. E93.C, no. 4, pp. 435–439, April 2010.

311. P. E. Hart, N. J. Nilsson, and B. Raphael, "A Formal Basis for the Heuristic Determination of Minimum Cost Paths," *IEEE Transactions on Systems Science and Cybernetics*, vol. 4, no. 2, pp. 100–107, July 1968.

312. N. Kito, K. Takagi, and N. Takagi, "Automatic Wire-Routing of SFQ Digital Circuits Considering Wire-Length Matching," *IEEE Transactions on Applied Superconductivity*, vol. 26, no. 3, pp. 1–5, January 2016.

313. C. H. Papadimitriou and K. Steiglitz, *Combinatorial Optimization: Algorithms and Complexity*, Dover, 1998.

314. N. Kito, K. Takagi, and N. Takagi, "A Fast Wire-Routing Method and an Automatic Layout Tool for RSFQ Digital Circuits Considering Wire-Length Matching," *IEEE Transactions on Applied Superconductivity*, vol. 28, no. 4, pp. 1–5, January 2018.

315. S. Kirkpatrick, C. D. Gelatt, and M. P. Vecchi, "Optimization by Simulated Annealing," *Science*, vol. 220, no. 4598, pp. 671–680, May 1983.

316. P. Cheng, K. Takagi, and T. Ho, "Multi-Terminal Routing with Length-Matching for Rapid Single Flux Quantum Circuits," *Proceedings of the IEEE/ACM International Conference on Computer-Aided Design*, pp. 1–6, January 2018.

317. Tim Edwards. Open Circuit Design. [Online]. Available: http://opencircuitdesign.com/qrouter/index.html

318. C. Y. Lee, "An Algorithm for Path Connections and Its Applications," *IRE Transactions on Electronic Computers*, vol. EC-10, no. 3, pp. 346–365, September 1961.

319. M. Kim, D. Lee, and I. L. Markov, "SimPL: An Effective Placement Algorithm," *IEEE Transactions on Computer-Aided Design of Integrated Circuits and Systems*, vol. 31, no. 1, pp. 50–60, January 2012.

320. T. Dejima, K. Takagi, and N. Takagi, "Placement and Routing Methods Based on Mixed Wiring of JTLs and PTLs for RSFQ Circuits," *Proceedings of the IEEE International Superconductive Electronics Conference*, pp. 1–3, July 2019.

321. S. Nath, K. English, A. Derrickson, A. Haslam, and J. F. McDonald, "An Automatic Placement and Routing Methodology for Asynchronous SFQ Circuit Design," *IEEE Transactions on Applied Superconductivity*, vol. 30, no. 3, pp. 1–10, April 2020.

322. Y. Murai, C. L. Ayala, N. Takeuchi, Y. Yamanashi, and N. Yoshikawa, "Development and Demonstration of Routing and Placement EDA Tools for Large-Scale Adiabatic Quantum-Flux-Parametron Circuits," *IEEE Transactions on Applied Superconductivity*, vol. 27, no. 6, pp. 1–9, June 2017.

323. T. Tanaka, C. L. Ayala, Q. Xu, R. Saito, and N. Yoshikawa, "Fabrication of Adiabatic Quantum-Flux-Parametron Integrated Circuits Using an Automatic Placement Tool Based on Genetic Algorithms," *IEEE Transactions on Applied Superconductivity*, vol. 29, no. 5, pp. 1–6, February 2019.

324. J. Lienig and K. Thulasiraman, "A Genetic Algorithm for Channel Routing in VLSI Circuits," *Evolutionary Computation*, vol. 1, no. 4, pp. 293–311, December 1993.

325. T. Yoshimura and E. S. Kuh, "Efficient Algorithms for Channel Routing," *IEEE Transactions on Computer-Aided Design of Integrated Circuits and Systems*, vol. 1, no. 1, pp. 25–35, January 1982.

326. K. Gaj, E. G. Friedman, and M. J. Feldman, "Timing of Multi-Gigahertz Rapid Single Flux Quantum Digital Circuits," *Journal of VLSI Signal Processing Systems for Signal, Image and Video Technology*, vol. 16, no. 2, pp. 247–276, June 1997.

327. M. El-Moursy and E. G. Friedman, *On-Chip Inductive Interconnect Design Methodologies*, VDM Verlag Dr. Muller Aktiengesellschaft & Company, 2009.

328. J. Rosenfeld and E. G. Friedman, "Design Methodology for Global Resonant H-Tree Clock Distribution Networks," *IEEE Transactions on Very Large Scale Integration (VLSI) Systems*, vol. 15, no. 2, pp. 135–148, February 2007.

329. M. E. Çelik and A. Bozbey, "Statistical Timing Analysis Tool for SFQ Cells (STATS)," *Proceedings of the IEEE International Superconductive Electronics Conference*, no. PA23, pp. 1–3, July 2013.

330. T. Kawaguchi, K. Takagi, and N. Takagi, "Static Timing Analysis of Rapid Single-Flux-Quantum Circuits," *Proceedings of the Workshop on Synthesis and System Integration of Mixed Information Technologies*, pp. 341–345, October 2016.

331. J. A. Delport and C. J. Fourie, "A Static Timing Analysis Tool for RSFQ and ERSFQ Superconducting Digital Circuit Applications," *IEEE Transactions on Applied Superconductivity*, vol. 28, no. 5, pp. 1–5, March 2018.

332. M. Dorojevets, "Architecture And Design Of An 8-Bit FLUX-1 Superconductor RSFQ Microprocessor," *International Journal of High Speed Electronics and Systems*, vol. 12, no. 2, pp. 521–529, June 2002.

333. S.-Y. Huang and K.-T. T. Cheng, *Formal Equivalence Checking and Design Debugging*, Springer Science & Business Media, 2012.

334. A. Fayyazi, S. Nazarian, and M. Pedram, "qEC: A Logical Equivalence Checking Framework Targeting SFQ Superconducting Circuits," *Proceedings of the IEEE International Superconductive Electronics Conference*, pp. 1–3, August 2019.

335. A. D. Wong, K. Su, H. Sun, A. Fayyazi, M. Pedram, and S. Nazarian, "VeriSFQ: A Semi-formal Verification Framework and Benchmark for Single Flux Quantum Technology," *Proceedings of the IEEE International Symposium on Quality Electronic Design*, pp. 224–230, March 2019.

336. I. Stotland, D. Shpagilev, and N. Starikovskaya, "UVM Based Approaches to Functional Verification of Communication Controllers of Microprocessor Systems," *Proceedings of the IEEE East-West Design & Test Symposium*, pp. 1–4, October 2016.

337. V. Adler, C.-H. Cheah, K. Gaj, D. K. Brock, and E. G. Friedman, "A Cadence-Based Design Environment for Single Flux Quantum Circuits," *IEEE Transactions on Applied Superconductivity*, vol. 7, no. 2, pp. 3294–3297, June 1997.

338. R. M. C. Roberts and C. J. Fourie, "Layout-Versus-Schematic Verification for Superconductive Integrated Circuits," *IEEE Transactions on Applied Superconductivity*, vol. 25, no. 3, pp. 1–5, June 2015.

339. I. P. Nevirkovets, S. E. Shafraniuk, O. Chernyashevskyy, D. T. Yohannes, O. A. Mukhanov, and J. B. Ketterson, "Investigation of Current Gain in Superconducting-Ferromagnetic Transistors With High-j_c Acceptor," *IEEE Transactions on Applied Superconductivity*, vol. 27, no. 4, pp. 1–4, June 2017.

340. I. P. Nevirkovets, S. E. Shafraniuk, O. Chernyashevskyy, D. T. Yohannes, O. A. Mukhanov, and J. B. Ketterson, "Critical Current Gain in High-jc Superconducting-Ferromagnetic Transistors," *IEEE Transactions on Applied Superconductivity*, vol. 26, no. 8, pp. 1–7, November 2016.

341. C. J. Fourie, C. L. Ayala, L. Schindler, T. Tanaka, and N. Yoshikawa, "Design and Characterization of Track Routing Architecture for RSFQ and AQFP Circuits in a Multilayer Process," *IEEE Transactions on Applied Superconductivity*, vol. 30, no. 6, pp. 1–9, 2020.

342. C. J. Fourie, C. Shawawreh, I. V. Vernik, and T. V. Filippov, "High-Accuracy InductEx Calibration Sets for MIT-LL SFQ4ee and SFQ5ee Processes," *IEEE Transactions on Applied Superconductivity*, vol. 27, no. 2, pp. 1–5, January 2017.

343. C. R. Paul, "A Simple SPICE Model for Coupled Transmission Lines," *Proceedings of the IEEE International Symposium on Electromagnetic Compatibility*, pp. 327–333, August 1988.

344. J. Zhang and E. G. Friedman, "Decoupling Technique and Crosstalk Analysis of Coupled RLC Interconnects," *Proceedings of the IEEE International Symposium on Circuits and Systems*, vol. II, pp. 521–524, May 2004.

345. J. Zhang and E. G. Friedman, "Mutual Inductance Modeling for Multiple RLC Interconnects with Application to Shield Insertion," *Proceedings of the IEEE International SOC Conference*, pp. 344–347, September 2004.

346. J. Zhang and E. G. Friedman, "Effect of Shield Insertion on Reducing Crosstalk Noise Between Coupled Interconnects," *Proceedings of the IEEE International Symposium on Circuits and Systems*, vol. II, pp. 529–532, May 2004.

347. J. Zhang and E. G. Friedman, "Crosstalk Modeling for Coupled RLC Interconnects with Application to Shield Insertion," *IEEE Transactions on Very Large Scale Integration (VLSI) Systems*, vol. 14, no. 6, p. 641–646, June 2006.

348. A. J. Kerman, E. A. Dauler, W. E. Keicher, J. K. W. Yang, K. K. Berggren, G. Gol'tsman, and B. Voronov, "Kinetic-Inductance-Limited Reset Time of Superconducting Nanowire Photon Counters," *Applied Physics Letters*, vol. 88, no. 11, p. 111116, March 2006.

349. Q. Liu, T. Van Duzer, X. Meng, S. R. Whiteley, K. Fujiwara, T. Tomida, K. Tokuda, and N. Yoshikawa, "Simulation and Measurements on a 64-kbit Hybrid Josephson-CMOS Memory," *IEEE Transactions on Applied Superconductivity*, vol. 15, no. 2, pp. 415–418, June 2005.

350. J. K. W. Yang, A. J. Kerman, E. A. Dauler, V. Anant, K. M. Rosfjord, and K. K. Berggren, "Modeling the Electrical and Thermal Response of Superconducting Nanowire Single-Photon Detectors," *IEEE Transactions on Applied Superconductivity*, vol. 17, no. 2, pp. 581–585, June 2007.

351. J. R. Clem and V. G. Kogan, "Kinetic Impedance and Depairing in Thin and Narrow Superconducting Films," *Physical Review B*, vol. 86, p. 174521, November 2012.

352. E. Testa, M. Soeken, L. G. Amarú, and G. De Micheli, "Logic Synthesis for Established and Emerging Computing," *Proceedings of the IEEE*, vol. 107, no. 1, pp. 165–184, January 2019.

353. A. L. Braun, "Large Fan-In RQL Gates," U.S. Patent No. 10,171,087, January 1, 2019.

354. A. Silver, R. Phillips, and R. Sandell, "High Speed Non-Latching SQUID Binary Ripple Counter," *IEEE Transactions on Magnetics*, vol. 21, no. 2, pp. 204–207, March 1985.

355. O. A. Mukhanov and A. F. Kirichenko, "A Superconductive High-Resolution Time-to-Digital Converter," *Proceedings of the IEEE International Superconductive Electronics Conference*, pp. 353–355, June 1999.

356. S. B. Kaplan, A. F. Kirichenko, O. A. Mukhanov, and S. Sarwana, "A Prescaler Circuit for a Superconductive Time-to-Digital Converter," *IEEE Transactions on Applied Superconductivity*, vol. 11, no. 1, pp. 513–516, March 2001.

357. L. Amarú, P. Gaillardon, S. Mitra, and G. De Micheli, "New Logic Synthesis as Nanotechnology Enabler," *Proceedings of the IEEE*, vol. 103, no. 11, pp. 2168–2195, November 2015.

358. P. Bunyk, K. Likharev, and D. Zinoviev, "RSFQ Technology: Physics and Devices," *International Journal of High Speed Electronics and Systems*, vol. 11, pp. 257–305, March 2001.

359. R. Cai, O. Chen, A. Ren, N. Liu, C. Ding, N. Yoshikawa, and Y. Wang, "A Majority Logic Synthesis Framework for Adiabatic Quantum-Flux-Parametron Superconducting Circuits," *Proceedings of the ACM Great Lakes Symposium on VLSI*, pp. 189–194, May 2019.

360. K. Jackman and C. J. Fourie, "Flux Trapping Analysis in Superconducting Circuits," *IEEE Transactions on Applied Superconductivity*, vol. 27, no. 4, pp. 1–5, June 2017.

361. V. K. Semenov and M. M. Khapaev, "How Moats Protect Superconductor Films From Flux Trapping," *IEEE Transactions on Applied Superconductivity*, vol. 26, no. 3, pp. 1–10, April 2016.

362. W. P. Burleson, M. Ciesielski, F. Klass, W. Liu, "Wave-Pipelining: a Tutorial and Research Survey," *IEEE Transactions on Very Large Scale Integration (VLSI) Systems*, vol. 6, no. 3, pp. 464–474, September 1998.

363. M. Dorojevets, C. L. Ayala, and A. K. Kasperek, "Data-Flow Microarchitecture for Wide Datapath RSFQ Processors: Design Study," *IEEE Transactions on Applied Superconductivity*, vol. 21, no. 3, pp. 787–791, June 2011.

364. S. S. Meher, J. Ravi, M. Celik, S. Miller, A. Sahu, A. Talalaevs, and A. Inamdar, "Superconductor Standard Cell Library for Advanced EDA Design," *IEEE Transactions on Applied Superconductivity*, vol. 31, no. 5, pp. 1–7, August 2021.

365. B. Dimov, M. Khabipov, D. Balashov, C. M. Brandt, F.-Im. Buchholz, J. Niemeyer, and F.H. Uhlmann, "Tuning of the RSFQ Gate Speed by Different Stewart-McCumber Parameters of the Josephson Junctions," *IEEE Transactions on Applied Superconductivity*, vol. 15, no. 2, pp. 284–287, June 2005.

366. C. J. Fourie, "Digital Superconducting Electronics Design Tools – Status and Roadmap," *IEEE Transactions on Applied Superconductivity*, vol. 28, no. 5, pp. 1–12, August 2018.

367. O. Mukhanov, A. F. Kirichenko, and D. E. Kirichenko, "Low Power Biasing Networks for Superconducting Integrated Circuits," U.S. Patent No. 9,853,645, December 26, 2017.

368. V. K. Semenov and Y. Polyakov, "Current Recycling: New Results," *IEEE Transactions on Applied Superconductivity*, vol. 29, no. 5, pp. 1–4, August 2019.

369. J. H. Kang and S. B. Kaplan, "Current Recycling and SFQ Signal Transfer in Large Scale RSFQ Circuits," *IEEE Transactions on Applied Superconductivity*, vol. 13, no. 2, pp. 547–550, June 2003.

370. M. W. Johnson, Q. P. Herr, D. J. Durand, and L. A. Abelson, "Differential SFQ Transmission Using either Inductive or Capacitive Coupling," *IEEE Transactions on Applied Superconductivity*, vol. 13, no. 2, pp. 507–510, June 2003.

371. K. Sano, T. Shimoda, Y. Abe, Y. Yamanashi, N. Yoshikawa, N. Zen, and M. Ohkubo, "Reduction of the Supply Current of Single-Flux-Quantum Time-to-Digital Converters by Current Recycling Techniques," *IEEE Transactions on Applied Superconductivity*, vol. 27, no. 4, pp. 1–5, June 2017.

372. V. K. Semenov and M. A. Voronova, "DC Voltage Multipliers: a Novel Application of Synchronization in Josephson Junction Arrays," *IEEE Transactions on Magnetics*, vol. 25, no. 2, pp. 1432–1435, March 1989.

373. T. V. Filippov, A. Sahu, S. Sarwana, D. Gupta, and V. K. Semenov, "Serially Biased Components for Digital-RF Receiver," *IEEE Transactions on Applied Superconductivity*, vol. 19, no. 3, pp. 580–584, June 2009.

374. G. Krylov and E. G. Friedman, "Partitioning of RSFQ Circuits for Current Recycling," *Proceedings of the IEEE Applied Superconductivity Conference*, November 2020.

375. N. K. Katam, B. Zhang, and M. Pedram, "Ground Plane Partitioning for Current Recycling of Superconducting Circuits," *Proceedings of the ACM/IEEE Design, Automation & Test in Europe Conference*, pp. 478–483, March 2020.

376. D. A. Papa and I. L. Markov, "Hypergraph Partitioning and Clustering," in *Handbook of Approximation Algorithms and Metaheuristics*. Chapman & Hall/CRC, 2007.

377. T. Manikas and G. R. Kane, "Partitioning Effects on Estimated Wire Length for Mixed Macro and Standard Cell Placement," *Proceedings of the ACM/IEEE International Workshop on Logic and Synthesis*, June 2002.

378. C. J. Alpert, A. E. Caldwell, A. B. Kahng, and I. L. Markov, "Hypergraph Partitioning with Fixed Vertices," *IEEE Transactions on Computer-Aided Design of Integrated Circuits and Systems*, vol. 19, no. 2, pp. 267–272, February 2000.

379. C. J. Alpert, T. Chan, D.-H. Huang, I. L. Markov, and K. Yan, "Quadratic Placement Revisited," *Proceedings of the ACM/IEEE Design Automation Conference*, pp. 752–757, June 1997.

380. R.-S. Tsay, E. S. Kuh, and C.-P. Hsu, "PROUD: a Sea-of-Gates Placement Algorithm," *IEEE Design & Test of Computers*, vol. 5, no. 6, pp. 44–56, December 1988.

381. B. W. Kernighan and S. Lin, "An Efficient Heuristic Procedure for Partitioning Graphs," *The Bell System Technical Journal*, vol. 49, no. 2, pp. 291–307, February 1970.

382. F. Brglez, D. Bryan, and K. Kozminski, "Combinational Profiles of Sequential Benchmark Circuits," *Proceedings of the IEEE International Symposium on Circuits and Systems*, vol. 3, pp. 1929–1934, May 1989.

383. F. M. Johannes, "Partitioning of VLSI Circuits and Systems," *Proceedings of the ACM/IEEE Design Automation Conference*, p. 83–87, June 1996.

384. B. Mathewson, "The Evolution of SOC Interconnect and How NOC Fits Within It," *Proceedings of the ACM/IEEE Design Automation Conference*, pp. 312–313, June 2010.

385. I. V. Vernik and D. Gupta, "Two-Phase 50 GHz On-Chip Long Josephson Junction Clock Source," *IEEE Transactions on Applied Superconductivity*, vol. 13, no. 2, pp. 587–590, June 2003.

386. Y. Zhang and D. Gupta, "Low-Jitter On-Chip Clock for RSFQ Circuit Applications," *Superconductor Science and Technology*, vol. 12, no. 11, p. 769, December 1999.

387. R. Ginosar, "Fourteen Ways to Fool Your Synchronizer," *Proceedings of the IEEE International Symposium on Asynchronous Circuits and Systems*, pp. 89–96, May 2003.

388. T. Jabbari, G. Krylov, S. Whiteley, J. Kawa, and E. G. Friedman, "Resonance Effects on SFQ Interconnect," *Proceedings of the Government Microcircuit Applications & Critical Technology Conference*, March 2020.

389. P. Horowitz and W. Hill, *The Art of Electronics*, Cambridge University Press, 1989.

390. S. Pasricha and N. Dutt, "ORB: An On-Chip Optical Ring Bus Communication Architecture for Multi-Processor Systems-on-Chip," *Proceedings of the ACM/IEEE Asia and South Pacific Design Automation Conference*, pp. 789–794, January 2008.

391. S. Narayana, V. Semenov, Y. Polyakov, V. Dotsenko, and S. Tolpygo, "Design and Testing of High-Speed Interconnects for Superconducting Multi-Chip Modules," *Superconductor Science and Technology*, vol. 25, no. 10, p. 105012, June 2012.

392. A. Joseph and H. Kerkhoff, "Towards Structural Testing of Superconductor Electronics," *Proceedings of the IEEE International Test Conference*, vol. 1, pp. 1182–1191, October 2003.
393. I. P. Litikov and O. A. Mukhanov, "Loop Self Testing of Josephson Junction Digital Structures," *Avtomatika i Vychislitelnaya Tekhnika [Soviet Automatics and Computers]*, no. 1, pp. 70–78, January 1988.
394. P. Girard, "Survey of Low-Power Testing of VLSI Circuits," *IEEE Design & Test of Computers*, vol. 19, no. 3, pp. 82–92, May/June 2002.
395. M. J. Y. Williams and J. B. Angell, "Enhancing Testability of Large-Scale Integrated Circuits via Test Points and Additional Logic," *IEEE Transactions on Computers*, vol. 22, no. 1, pp. 46–60, January 1973.
396. L.-T. Wang, C.-W. Wu, and X. Wen, *VLSI Test Principles and Architectures: Design for Testability*, Elsevier, 2006.
397. C.-C. Lin, M. Marek-Sadowska, K.-T. Cheng, and M.-C. Lee, "Test-Point Insertion: Scan Paths Through Functional Logic," *IEEE Transactions on Computer-Aided Design of Integrated Circuits and Systems*, vol. 17, no. 9, pp. 838–851, September 1998.
398. D. K. Brock, E. K. Track, and J. M. Rowell, "Superconductor ICs: the 100-GHz Second Generation," *IEEE Spectrum*, vol. 37, no. 12, pp. 40–46, December 2000.
399. T. Fulton, "Externally Shunted Josephson Junctions," *Physical Review B*, vol. 7, no. 3, p. 1189, February 1973.
400. J. L. Neves and E. G. Friedman, "Design Methodology for Synthesizing Clock Distribution Networks Exploiting Nonzero Localized Clock Skew," *IEEE Transactions on Very Large Scale Integration (VLSI) Systems*, vol. 4, no. 2, pp. 286–291, June 1996.
401. V. K. Semenov, "Digital SQUIDs: New Definitions and Results," *IEEE Transactions on Applied Superconductivity*, vol. 13, no. 2, pp. 747–750, June 2003.
402. Y. C. Kim and K. K. Saluja, "Sequential Test Generators: Past, Present and Future," *Integration, The VLSI Journal*, vol. 26, no. 1, pp. 41–54, December 1998.

Index

© Springer Nature Switzerland AG 2022
G. Krylov, E. G. Friedman, *Single Flux Quantum Integrated Circuit Design*,
https://doi.org/10.1007/978-3-030-76885-0

Printed in the United States
by Baker & Taylor Publisher Services